Windmills of Norfolk

John Ling

AMBERLEY

First published 2015

Amberley Publishing
The Hill, Stroud, Gloucestershire, GL5 4EP
www.amberley-books.com

Copyright © John Ling, 2015

The right of John Ling to be identified as the Author
of this work has been asserted in accordance with
the Copyrights, Designs and Patents Act 1988.

ISBN 978 1 4456 5377 8 (print)
ISBN 978 1 4456 5378 5 (ebook)

British Library Cataloguing in Publication Data.
A catalogue record for this book is available from
the British Library.

Typesetting by Amberley Publishing.
Printed in Great Britain.

Contents

Introduction

The large county of Norfolk is a haven for windmill enthusiasts. From giant tower mills to tiny wooden structures all windmill life is here. Under the wide Norfolk skies the mills stand like sentinels, often visible for miles around. They are now viewed through a haze of nostalgia and with more than a hint of romance, but their role today as decorative tourist attractions would probably mystify the millers of a bygone age who slaved long and hard to eke out a living. Over 400 corn mills were working in Norfolk around 1850 but this figure had dropped to less than 100 by the outbreak of the First World War. Croxton Mill near Fulmodeston was the final corn tower mill to be built in Norfolk in 1885. The last to work commercially was Billingford Mill, located in the south of the county close to the border with Suffolk, which continued to use wind power until 1956. Though an awful lot have been completely or partially demolished over the years, or are now just empty shells, many fine examples remain. In recent decades councils, windmill trusts and private owners have restored a number of mills to a virtually 'as new' condition. Often this transformation stops short of actually getting the sails turning again but a few have been brought back to full working order. Traditionally, windmills were referred to as 'she' but whether that is politically correct in the twenty-first century I don't know.

Windmills of Norfolk takes an in-depth look at some of the county's tallest, largest, finest and most famous corn and drainage mills. It also casts an eye over smaller, lesser known and more obscure examples. From the record-breaking giants of yesteryear to much more modest structures, there is hopefully something of interest for all lovers of windmills and also those new to the subject. It does not pretend to be a highly technical account of how these machines operate but is written from the personal viewpoint of a lifelong windmill enthusiast. It is intended to inform and entertain in equal measure.

During the course of my research, I have made every effort to unearth accurate and reliable information. Inevitably, even reputable sources sometimes disagree with each other over various points including a mill's height, sail or tower diameter, number of floors, date of erection or closure, building costs etc. Where considered appropriate, these discrepancies are mentioned in the text. Location information is included for those who may wish to visit surviving mills and it is clearly stated which ones are open to the public. Not all of these are open all year round and it is worth checking before travelling. Opening dates and times are based on information available at the time of writing. Of those that are not open to the public, most can be viewed from the roadside without the need to trespass on private property. Some are inaccessible by road and can only be reached by river, rail or on foot, and this is explained in the text.

Historical and Technical Information

Long before arriving in Norfolk the origins of the windmill can be traced back to ancient Persia (now Iran). The first corn mills appeared there as early as the seventh century before eventually spreading to Europe. A mill was recorded near Hull in 1185 and the first examples in Norfolk appeared during the thirteenth century. From tiny, very simple wooden post mills (so called because they were constructed around a central oak post extending from ground level towards the top of the mill, which was originally left open to the elements), the windmill eventually grew into a highly sophisticated machine. The post mill evolved into a fully enclosed and larger structure with the whole wooden body or buck being rotated in accordance with the wind direction. This sat on a brick roundhouse that remained static.

The sixteenth century gave birth to the smock mill, which consisted of a fixed wooden building on a brick or stone base, with a rotating cap. For the first time the sails could be moved without turning the whole body of the mill. The base was often little more than a foundation on which to fix the main timber body, though in the case of Briningham Smock Mill the brickwork was three storeys and about 29 feet (8.84 metres) tall.

It was the advent of the tower mill, entirely constructed from bricks, that enabled much larger structures to be built, and advances in technology eventually led to the giants which appeared in the late eighteenth and nineteenth centuries. Taller and wider towers and larger sails soon became the norm with some corn mills being equipped with five, six or even eight sails rather than the usual four. In exceptional cases some grew to well over a hundred feet (30.5 metres) tall. Ironically, the windmill reached its peak of technical perfection at almost the same time as its sudden decline began. There is little doubt that ever larger mills would have been constructed had advances in other technologies such as steam engines and, later, electricity not sounded their death knell.

Windmills equipped with more than the standard complement of four sails are often known as multi-sailed mills or multi-sailers. More common in neighbouring Lincolnshire, that county still has examples with five, six or eight sails, including the only surviving British eight-sailed mill at Heckington. Norfolk now has none but once had half a dozen known six-sailers. These included: Balsam Fields and Orange Farm Mills in Terrington St Clement; West Walton Ingleborough and West Walton Highway Mills; Ringstead Mill and Sculthorpe Mill. Of these six the remains of four still exist in one form or another but today they cannot muster a single sail between them. Mills in

Old Buckenham and Diss both had eight sails before being converted to conventional four-sailers. The tallest ever eight-sailed mill in Britain was Leach's Mill in Wisbech, just over the border in Cambridgeshire.

The normal method of measuring the height of a windmill is from the base to the top of the tower but not including the sails. Often the measurement includes the wooden cap atop the tower but sometimes it is only to the top of the brickwork (known as the curb), which can lead to confusion. On a large mill the cap could add an additional 15 feet (4.57 metres) or more to the height of the tower. The number of storeys obviously contributes greatly to the overall height though the distance between floors varies widely from one mill to another. It is possible for a mill with, for instance, six floors to be taller than one with seven or eight. Eleven appears to be the highest number of storeys definitely attributed to any known mill, though an unsubstantiated claim of thirteen has been made. Usually only the number of floors in the tower itself is included but in some cases the cap is counted as a top level. In some instances the claimed height is to the tip of the topmost sail, which if not clearly stated could be regarded as cheating! The tallest windmill may not necessarily be the largest, as the latter term can refer to the overall diameter of the sails rather than the height of the structure. In addition, some mills lay claim to having the widest diameter tower or the largest cap. Again, Norfolk mills are well represented in these categories.

The working tower windmill was not just a building but a massive and intricate piece of specialised machinery. In simple terms the sails were connected by means of gears and other machinery to millstones which ground grain into flour. The fantail, also called fan or fly (a circular device made up of small blades and fixed to the fan stage at the rear of the wooden cap), automatically steered the cap assembly round so that the sails faced the wind. The traditional corn mill was once extremely common, grinding wheat or other grains in addition to corn for the local community and beyond by means of environmentally friendly wind power. Many mills later went over to grinding animal feed, particularly after the First World War, and the same principal could be used to grind various other substances.

During the eighteenth century, drainage mills or windpumps began to appear in Britain alongside the corn mills. Instead of grinding with stones these were designed to lift water from low-lying areas and were useful in providing additional land suitable for growing crops or grazing cattle. Water was raised by means of a scoop wheel either inside or outside the mill and connected to the sails by a shaft and smaller wheels inside, though other methods such as a turbine pump or plunger pump were sometimes used instead. The windpump became popular in the Norfolk Broads and the Fens but was often a small and very basic structure not unlike very early corn mills, sometimes consisting of a skeleton of timbers open to the elements which supported tiny sails. Many later examples had brick towers but few drainage mills grew as large as their corn-grinding cousins. A notable exception to that rule is the superb Berney Arms Mill near Reedham, which is unique in having originally been built to grind cement clinker before being converted to draining marshland.

By today's standards, when it often takes many years and an extraordinary amount of money to restore a windmill to something approaching its former glory, it is amazing how swiftly these giant structures were erected without modern aids such as massive cranes and computer models. This was, of course, well before the age of stringent health and safety regulations, which if in force at that time would no doubt have seen the immediate closure of many mills. Most early ones had sails which almost scraped the ground, making it hazardous to enter or leave when a strong wind was blowing. Sadly, several millers met a grisly fate either trapped in the internal machinery or by literally losing their heads to a passing sail! Following William Cubitt's invention of the patent self-regulating sail in 1807, it was no longer necessary for the miller to set the canvas by hand as on the early common sails. This meant that the sails could now be placed well away from the ground and resulted in a lot of mills being raised or hained sometimes two or three times during their working lives. In addition to ease of operation and safety, the power of the wind could be harnessed more effectively if the sails were higher. Many windmills were built on high ground to make the most of the prevailing wind, often dominating their surroundings despite being of relatively modest proportions.

Chapter 1

Surviving Tower Corn Mills

Norfolk still has many surviving tower corn mills, the condition of which range from working order to derelict and everything in between. Several have been converted to residential use and a number have been truncated. This chapter looks at some of the most famous and lesser known examples.

THE NORFOLK BROADS AND NORWICH AREA

Sutton Mill near Stalham
Location: Mill Road, Sutton, NR12 9RZ. Approx. 0.5 mile east of Sutton village, off A149, near Stalham. Privately owned. Not open to the public at present.

Left and opposite: Sutton Mill, October 1937. (J. Middleton)

For many years Sutton Mill was listed in *The Guinness Book of Records* as Britain's tallest surviving windmill, though its actual height was not documented. It has now lost its national record to Lincolnshire's Moulton Mill but can still claim to be the tallest in East Anglia and possibly the second or third tallest in the country. Although not built on such a massive scale as some of its Norfolk contemporaries (see Chapter 5, 'The Lost Giants') it is nevertheless a large and imposing mill standing 67 feet 6 inches (20.57 metres) tall to the curb of its nine-storey tower and just short of 80 feet (24.38 metres) to the top of the cap. The painted red-brick tower has an impressive 33 feet (10 metres) diameter at ground level and 16 feet (4.88 metres) at the top. Its original four sails had 216 shutters and spanned 73 feet (22.25 metres) in diameter. The mill was run by the Worts family throughout its working life before being struck by lightning in 1940. The previously unpublished period photographs show the sails intact less than three years before they stopped turning. During its productive years it was painted white and housed five pairs of millstones. Four pairs (each measuring 4 feet 6 inches (1.37 metres) in diameter) survive on the fifth floor but a pair on the second floor is no longer in place. The tower still has a reefing stage on the fifth floor and previously had a gallery at the base of the typically Norfolk boat-shaped cap. Standing in a rural location just outside the village of Sutton near Stalham, the mill is a landmark for miles around. There has been some confusion over its date of construction but the current mill probably dates from between 1859 and 1862 and replaced an earlier eight storey one erected around 1789 and destroyed by fire between 1857 and 1861. The nearby mill house is said to date from the time of the previous mill.

Sutton Mill in need of restoration, October 2012.

After being allowed to deteriorate following the lightning strike, Sutton Windmill was rescued from dereliction by the Nunn family in 1975. They carried out extensive repairs over the course of nearly thirty years and opened it to the public. A large Broads Museum was later added in a building on the same site, displaying all manner of artefacts from the past. Sadly, lack of funds prevented the Nunns from restoring the mill to full working order, which from the outset was their declared aim. The remnants of the fire-damaged sails were removed and replaced with four new stocks (the main timbers to which sails are fixed) and for a while new sails were also fitted. Mr Nunn advertised the mill and Broads Museum for sale in the early twenty-first century and after being purchased by new owners the site reopened to the public following a period of closure in April 2007. Unfortunately, it closed again on 26 September 2008 due to the deteriorating state of the building. Since then it has remained closed and the estimated cost of a full restoration is at least £500,000. Over a thousand lots from the mill's museum were offered for sale at onsite auctions on 11 and 12 September 2012. Work to remove the cap and sail stocks and fit a temporary flat roof finally began in early September 2014, following severe storm damage the previous winter.

Sutton Mill has a tarnished grandeur and a slightly ethereal quality. It is one of my favourite mills and reminds me of a bruised and battered heavyweight boxer well past his peak who refuses to throw in the towel. When I first encountered it a few decades ago both the mill and mill house were in a very sorry state and looked rather spooky as dusk approached. The former still carried the charred remains of its sails and the

Sutton Mill repairs begin,
September 2014.

latter was derelict and in danger of being overrun by its long neglected garden. The house has since fared better than the wounded giant that towers above it, having been fully restored and run for many years as a residential home, though it is no longer used for that purpose. Speaking as one of the many visitors who climbed to the top of the tower and stood outside on the now vanished narrow gallery to savour the impressive views, Sutton Mill certainly deserves to be saved for future generations. Hopefully, the dream of returning Norfolk's tallest surviving windmill to full working order may still be realised but it would require a massive injection of cash and commitment similar to that lavished on the mill which took its title as Britain's tallest.

Hickling Mill near Stalham

Location: Approx. 2 miles southeast of Stalham between Hickling Heath and Hickling Green, NR12. Privately owned and not open to the public.

Standing close to the road that runs between Hickling Heath and Hickling Green, around 2 miles from Stalham, is one of the county's tallest but lesser known windmills. If Hickling Mill had retained its sails (removed in 1907) it would no doubt generate more interest but the eight-storey tower is worthy of attention on several fronts. Of the remaining Norfolk mills only Sutton, which it faces across a couple of miles of flat arable farmland, has more floors. Hickling Mill is about 61 feet (18.59 metres) tall to the curb of the tower and 71 feet (21.64 metres) to the ridge of its cap. It is of similar height to Berney Arms Mill (see Chapter 2), though that has one floor fewer and a full

Hickling Mill, October 2012.

set of sails. The tower has an internal diameter of 25 feet (7.62 metres) at base level with brickwork 2 feet 6 inches (0.76 metres) thick. It has no reefing stage but has a gallery on the cap. One of the most unusual features is that the windows on each floor are all set one above the other, and also the sheer number of them. Most tall windmills have offset windows as this helps the rigidity of the structure, but at Hickling all is not as it first appears. From the first to seventh floors, thirteen out of twenty-eight windows are said to be dummies, visible only from the outside, which shows the effort put into creating an aesthetically pleasing appearance.

Hickling Mill is thought to have been built in 1818 or slightly earlier for William Crowe. Israel Royal Garrett, a miller, farmer, merchant and baker, was at the mill from around 1845 to 1883. The last miller was Henry Wright, who by 1904 had ceased milling. Thirty years later it was purchased by the Forbes family, who still own it today. Major work was carried out in 1989/90 by Norfolk millwright Richard Seago. This included rebuilding the cap and fan stage but did not extend as far as refitting the sails and fantail. This seems a pity as the mill deserves to be restored to its former glory given its height and unusual architectural features. From a financial and maintenance viewpoint, however, it makes sound sense. The mill is not accessible to the public but can be clearly seen from the road.

Ingham Mill near Stalham

Location: Mill Farm, Mill Road, Ingham, NR12 9TD. Off B1159, approx. 1.5 miles northeast of Stalham. Privately owned and not open to the public.

A tower corn mill is thought to have existed on this site since about 1763. Before that, an even earlier post mill was toppled during a gale. The present tower mill was

Left: Ingham Mill, October 1937 (J. Middleton). *Right*: September 2012.

built around 1872 after the demolition of its predecessor. The five-storey tower now stands 49 feet (14.93 metres) tall but in its working days was 42 feet (12.80 metres) to the curb. It had four double-shuttered patent sails driving three pairs of millstones, a familiar Norfolk boat-shaped cap complete with gallery and a fantail with six blades. George Matthews worked the old tower mill from 1836 and was listed as owner in 1842. He may have begun construction of its replacement before his death in April 1872. Among the other millers who plied their trade at Ingham were: Henry Cook (*c.* 1878–79), George Robert Gladden (*c.* 1883–1916) and Norman Goffin (*c.* 1922–37). The previously unpublished period photograph clearly shows that in October 1937 the cap, stocks and one pair of sails were still in place, though the latter's shutters were missing or damaged.

Between 1939 and 1947 the Royal Observer Corps took over the mill. It was during their occupancy that the cap and sails were removed 7 feet (2.13 metres) of upright brickwork was added. From the outside the mill today still looks pretty much as it did then. That said, it shows definite signs of wear and tear, as can be seen in the second photograph taken almost exactly seventy-five years after the first. It is on private land and not open to the public but stands close to the road.

Potter Heigham Mill

Location: Bottom of Mill Road, Potter Heigham, NR29 5HY. Off A1062 heading east from Hoveton; or turn off A149 at Potter Heigham Bridge. Privately owned and not open to the public.

This imposing tower mill in the east of the county was erected in 1849 by Suffolk millwrights Martins, for Simon Boyce. It replaced an earlier post mill and stood six

Potter Heigham Mill, July 2012.

storeys tall plus the cap. It had four double-shuttered patent sails which drove two pairs of French Burr millstones. The tower's internal diameter is 23 feet (7.01 metres) at the base with 1.5 feet (0.46 metres) thick walls. It stands around 54 feet (16.46 metres) tall to the curb and the original cap probably added another 10 feet (3.05 metres). At some point the top floor windows were bricked up. A Ransomes 10 hp steam engine was generating additional power by 1900 but the mill continued to work by wind until about 1928 and still retained its cap, stocks, a pair of sails and the remnants of the fantail for several more years. These had all been removed by late 1953, when a conical galvanised iron roof was fitted. The mill last worked using electricity till at least the end of the 1940s and some of the original internal machinery is still *in situ*.

Since the 1950s Potter Heigham Mill has changed hands on several occasions. In 2009 the owners launched an appeal to raise £1,000 in order to carry out essential repairs to the brickwork and roof. The Grade-II listed tower is one of the largest left in the area.

Ludham – How Hill Mill
Location: How Hill, Ludham, NR29 5PG. Opposite How Hill House, close to public car park of How Hill Nature Reserve, on narrow lane off A1062, approx. 1.5 miles west of Ludham. Also accessible from the east bank of the River Ant. Privately owned and available for holiday lets.

A date stone inscribed 'W.S.P. 1825' is incorporated into the tower of How Hill Mill. The initials refer to William Sherwood Page, who built the mill and was still the owner in 1851. Throughout his tenure and beyond a number of different millers worked at the site. The mill and a 190-acre farm was still in ownership of the Page family in 1888, when it was advertised for let. During its working life the mill had a boat-shaped cap, four double-shuttered sails and a fantail. By 1890 it was showing obvious signs of neglect but the remnants of the sails and fantail were still in position. After the How Hill estate was purchased in 1902 by architect Edward Thomas Boardman, the mill tower was adapted to house water tanks to help irrigate the farm. A new flat roof replaced the decayed cap and the former mill continued to fulfil this role for several decades. Mr Boardman stayed at How Hill until the mid-1960s. The water tanks and all internal machinery were removed prior to its conversion to residential use around 1971.

The five-storey tower is notable for its unusually pronounced batter (the amount of slope of the walls between the bottom and top of the structure). From a base diameter of 21.5 feet (6.55 metres) it tapers to just 10 feet (3.05 metres) at the curb. Two other corn mills once existed in the Ludham area but of these no trace remains. The present-day visitor to How Hill can still observe the restored Turf Fen Drainage Mill (see Chapter 2) and enjoy close-up views of two small timber windpumps beside the River Ant (see Chapter 4).

How Hill Mill, June 2013.

Norwich – Lakenham Peafield Mill

Location: Eleanor Road, Norwich, NR1 2RE. Converted to residential use.

The sole survivor and one of the tallest of many mills which once worked within the boundaries of the fine city of Norwich, the eight-storey black tower of Peafield Mill remains a local landmark. It has parted company with its sails, fantail and original cap and been converted to residential use, though for safety reasons only the first, second and third floors were converted. It has been claimed that the sails had a combined span of nearly 100 feet (30.48 metres) and the tower was 80 feet (24.38 meters) tall to the curb plus the cap. This would make it taller than the nine-storey tower of Sutton Mill – credited as Norfolk's highest since 1920 – but I am unable to verify these figures.

As its name suggests, Peafield Mill was built in open countryside near Norwich in 1824. Fifty years later it was still fairly isolated but was soon swallowed up by the increasing urbanisation of its surroundings. Today it is part of one of the city's busiest residential districts. Also known as Spratt's Mill and the Black Mill at various stages of its life, it was constructed at a cost of over £2,000 by Henry Lock, who also built the Lakenham Bracondale Mill (see Chapter 5). The original miller, Thomas Spratt, was replaced in 1833 by Charles Clare, who himself moved on five years later to take charge of the mighty Bixley Mill (see Chapter 5). His successor at Peafield was James Faulke before the mill was sold at auction on 19 August 1845, at the Royal Hotel, Norwich. At that time it was described as having nine storeys, the cap floor obviously being counted as the top level. The purchaser was Barnabas Feltham, who passed away about ten years later and was superseded by his son, also called Barnabas.

Lakenham Peafield Mill, Norwich, January 2012.

The Mill temporarily lost its fantail in a gale in early February 1884, part of which caused damage to a cottage some distance away. After being bought by Charles Edward Woodrow in 1905 the mill was part of his company's milling business for the next sixty years. Following the loss of its sails around 1914/5, the mill continued to operate by electric power. The original cap was still in place in the 1930s but was probably taken down soon after. The mill was acquired by the Norwich Corporation in 1970. In a perfect world Peafield Mill would have been restored to full working order. In reality, the change of use saved it from going the same way as all the other Norwich windmills and ensured its continued existence. As the last one still standing it serves as a reminder of the city's proud milling heritage.

Frettenham Mill near Norwich

Location: Mill Farm, Mill Road, Frettenham, NR12 7LQ. Approx. 6 miles north of Norwich. Privately owned and not open to the public.

Frettenham Mill was built between 1870 and 1880 for Joshua Harper, who died in 1891. Alfred Herne purchased it at auction and ran it until around the turn of the century. By 1910 the sails and fantail had been taken down, so the mill's working life was relatively short. In 1937 it was said to still have a reefing stage on the second floor of the five-storey tower – which stands 47 feet (14.32 metres) tall to the top of the brickwork – along with the cap and fantail frames. After being derelict for many years the mill was finally renovated and turned into residential accommodation by 2004. A new cap had been installed by the year 2000 and a replacement reefing gallery was fitted later. Some of the internal machinery and gearing was left in place but the work did not include replacing the sails. The mill, along with converted outbuildings, was put on the property market in 2008.

Frettenham village sign incorporates one of the original French Burr millstones as part of the base. It was erected in 1981 and the stone was donated by the late Mr W. D. Field of Mill Farm.

Frettenham Mill, March 2012.

Horsford – St Helen's Mill near Norwich

Location: Mill Lane, Horsford, NR10 3ES. Just off B1149 behind public house, approx. 6 miles northwest of Norwich. Privately owned and not open to the public.

Known to some as St Helen's Tower Mill and to others simply as Horsford Mill, this relatively modest structure was erected around 1860–65 by Elijah Punt and replaced a smock mill dating back to about 1820. Despite having five floors the tower was said to be only around 36 feet (10.97 metres) tall to the top of the brickwork. Located in an area of Horsford known as St Helena, it was equipped with a boat-shaped cap, four double-shuttered sails and a fantail, but these were all taken down long ago. The first miller was Charles Punt, who had latterly worked the smock mill. It was sold at auction in 1868 to William Howe and from 1883 became associated with the well-known Woodrow family of millers and bakers. Edward Woodrow, who had previously operated Billingford and Scole tower mills, began working at Horsford in that year as a tenant miller and bought the mill in 1889. He was later assisted by his sons Charles (who would later acquire Peafield Mill at Lakenham), Frederick and Hedley. The mill stopped working in the early 1920s, by which time Hedley Woodrow was in charge at Horsford. The tower still carried an advertisement for Hovis in the early 1960s. In 1978 the dust floor was leaded over to provide a flat protective roof. The windshaft was finally taken down in September 1982.

Horsford Mill was sold in 2005 and restoration began in 2007. The work also saw the building of a large extension and partial conversion of the tower to residential use. A new cap was fitted in 2015.

St Helen's Mill, Horsford, July 2015.

Aylsham – Cawston Road Mill

Location: Just off B1145, Cawston Road, Aylsham, NR11 6NB. Approx. 12 miles north of Norwich. Privately owned and run as a B&B business.

Against the odds, the remains of a once tall windmill still stand close to Cawston Road, Aylsham. In its working days a prominent landmark on the edge of the town, it is now surrounded by other buildings. Originally seven storeys high plus cap, the top floor was reportedly lost during an abortive attempt at demolishing the tower after the cap and sails were blown off around 1920. Apparently the brickwork proved remarkably resilient and the truncated building was reprieved for posterity. This is very similar to the story of Stoke Ferry Mill, which is also said to have survived the ultimate fate!

Cawston Road Mill was erected in 1826 for Henry Soame, who died seven years later. His son, George, inherited the mill and later leased it to John Neech, who became bankrupt in 1860. Soame took it over again but put it up for sale at auction in 1865 after defaulting on a mortgage. Miller James Davison bought Cawston Road Mill in 1872 for the princely sum of £315. After ceasing work and standing empty for decades, the tower was finally renovated and converted to residential use in 1998. Now an unusual bed and breakfast establishment, The Old Windmill – as it is now known – has no cap but retains its distinctive uneven top as evidence of that botched demolition attempt.

Another tall corn mill also existed in Aylsham from around 1826 to 1941, when it was demolished. Buttlands Lane Mill was also operated by James Davison and originally had two common and two patent sails, driving two pairs of stones. All four sails were of the patent type by the late 1850s. Unfortunately, they parted company with the rest of the mill during a gale in March 1895, and were not refitted. An unsubstantiated height of 85 feet (25.9 metres) has been claimed but whether this figure is reliable remains uncertain.

Cawston Road Mill, Aylsham, March 2012.

NORTH NORFOLK

Burnham Overy – Savory's Mill

Location: Tower Road, Burnham Overy, PE31 8JB. On A149 coast road, approx. 0.5 mile west of Burnham Overy and 1 mile north of Burnham Market. Owned by the National Trust and available to rent as holiday accommodation.

Standing proudly beside the coast road in North Norfolk and set in one of the county's most beautiful locations, the large six-storey black tower mill at Burnham Overy was built in 1816 by Edmund Savory. His initials and the date of construction still adorn the tower above the first floor level. He operated it until 1827, when it passed to his son John. It continued in the same family until 1900, having been taken over by John Savory Junior in 1863. The last two millers were Sidney Dewing (1900–10) and Sidney Everitt (1910–21). The mill stopped working after being tail-winded in 1914 and was purchased by Mr H. C. Hughes in 1926. Most of the machinery was removed – apart from some in the cap – before conversion to holiday accommodation. At its base the tower is around 24 feet (7.32 meters) in diameter with 2 feet (0.61 meters) thick brickwork. It originally had three pairs of millstones and the sails reportedly spanned 81 feet (24.69 meters) in diameter. A new fixed cap, stocks and skeleton sails were put in place in 1957 by the well-known Lincolnshire millwrights R. Thompson & Son of Alford, who also carried out additional work. The newly renovated mill was presented to the National Trust by Mr Hughes the following year. The exterior was retarred in 1981 and extensive work was undertaken inside the tower to bring it in line with fire regulations. Another set of sails and stocks were put in place in July 1985, the previous ones having been removed in 1983. The fantail was also renovated. The total cost of the work carried out between 1981 and 1985 was said to be more than £50,000.

Savory's Mill, Burnham Overy, July 2009.

Savory's Mill, Burnham Overy, July 2009.

Savory's Mill still retains its four replacement sails and fantail in addition to the ogee-shaped cap complete with gallery and blue finial, and is immaculate in appearance. One of the region's best known and most photographed windmills, it is not open to the public but is available to rent through the National Trust throughout the year. As such, it is one of the county's grandest and most desirable holiday 'cottages', with coastal views from its upper floors. The reefing stage on the first floor now serves as a veranda for the paying guests.

Cley Mill, Cley-next-the-Sea near Holt
Location: The Quay, Cley-next-the-Sea, NR25 7RP. On A149 east of Holt. Privately owned and run as a guesthouse.

Although not exceptionally tall, few windmills dominate their surrounding countryside for miles around like the substantial red-brick one at Cley-next-the-Sea in North Norfolk. Situated approximately 14 miles from Savory's Mill at Burnham Overy, Cley Mill's location is arguably even more breath-taking. It is one of Britain's most famous windmills, due in part to appearances in films and on television, and has been described as 'probably the most painted and photographed mill in the country.' It was seen in the MGM film *Conspirator* in 1949 and in *Conflict of Wings* five years later. Television appearances include *The Ruth Rendell Mysteries*, a series featuring

Cley Mill, July 2009.

naturalist David Bellamy, and a long-running BBC 1 trailer in which it 'starred'
opposite a hot air balloon passing overhead. On this occasion the sails could be seen
apparently revolving, though both they and the cap were reportedly fixed to prevent
them turning during the 1920s. More recently, like Berney Arms Mill (see Chapter 2)
and several other Norfolk landmarks, it graced the opening sequences of Anglia TV's
drama, *Kingdom*, starring Stephen Fry.

Back in the Middle Ages, when Cley really was next-the-Sea, there was a flourishing
port here, but after reclamation of land in the seventeenth century the town found
itself some distance from the waves. Following a fierce fire which consumed much
of the old town, new buildings and wharves were constructed to form the basis of
the picturesque village which exists today. Cley Mill is thought to have been built on
or close to the site of an earlier post mill on the quay, sometime between 1810 and

Cley Mill, July 2009.

1819. Having survived the closure of the port in 1884, the mill continued to grind corn using three pairs of millstones before it finally ceased work by 1917. Four years later it was purchased by Mrs Sarah Maria Wilson and became the first of several mills on the Norfolk coast to be converted to residential use. Others include the aforementioned Savory's Mill at Burnham Overy and Weybourne Mill, which has a full-sized house attached to its tower. Most of the machinery including the millstones was removed when Cley Mill became a holiday home and the warehouses were turned into boat sheds and stables. Lt. Col. Hubert Blount, who inherited the mill from his grandmother in 1934, replaced the sails in 1960. Further renovations took place in 1983, when it became an upmarket guest house. In 1986 it was announced that it would cost around £45,000 to replace the sails, fantail and gallery. On 5 August 1987, a new fantail and frame made by millwright John Lawn were lifted into place by a crane. The mill later passed to Charles and Jane Blount (parents of the successful singer-songwriter known professionally as James Blunt). In 2006 it changed hands again.

Today, Cley Mill retains its four shutterless sails and distinctive polygonal dome-shaped cap complete with finial and gallery. There is still a reefing stage on the second floor of the wide five-storey tower. The photographs reveal that during the summer of 2009 the fantail was absent but this has since been replaced. At the time of writing the sails are once more in need of repair or replacement.

Cley Mill July, 2009.

Weybourne Mill, near Sheringham

Location: Sheringham Road, Weybourne, NR25 7EY. On north side of A149 coast road between Sheringham and Cley, approx. 3 miles west of Sheringham and 5.3 miles east of Cley. Privately owned and not open to the public.

Along with the famous windmills at Cley and Burnham Overy (Savory's), this prominent landmark will be familiar to travellers on the A149 coast road. Like them, Weybourne Mill looks externally complete but had its machinery removed during conversion to residential accommodation.

The mill was built in 1850 and worked till about 1916. Between these dates, several different millers passed through its doors beginning with Daniel Brett, who stayed until 1858. John Dawson then took charge before being made bankrupt around 1869. Others came and went before Thomas Youngman arrived in the mid-1890s. He was still there when the mill ceased operating during the First World War, when it served as a military lookout post. During its productive life the four original double-shuttered sails drove three pairs of stones on the second floor, where a reefing stage once encircled the five-storey tower. The approximate height to the curb was a fairly modest 42 feet (12.8 metres) and a traditional boat-shaped cap probably added another 10 feet (3.05 metres). A fantail with eight blades completed the picture.

A granary was attached to the tower and an internal door enabled easy access between the two. When the mill was renovated in the mid-1920s the granary's roofline was made higher as part of its conversion to a dwelling. The main timber post from Weybourne Post Mill was reused in the house and a 'miller's cash box' was inserted

Weybourne Mill, November 2012.

into the woodwork. The new owner was an architect named John Sydney Brocklesby (1879–1955), who designed the conversion and removed the mill's sails. A few years later he was said to be interested in refitting the sails with the intention of using wind power to produce electricity for his own needs, but this ambitious plan failed to reach fruition. It would be another forty years before Weybourne Mill received a new set of sails, but they never moved an inch. The replacements fitted in 1969 were the skeleton type (without shutters) and they were connected to the original windshaft – the only piece of machinery still *in situ* – and fixed in place to prevent them from turning. A new black cap had replaced the original white-painted one a year earlier. The mill was now the property of Mr W. Boby, who became the proud owner in 1967. Further work took place in the early 1970s, helped by grants from Norfolk Windmills Trust. The new sails survived a lightning strike in January 1978, the only damage being a ruptured water pipe. The mill and mill house were sold for a reported £125,000 in 1982. Four years later, millwright John Lawn carried out repairs to a sail. The mill was pictured without sails in April 2008, but they were soon back in position.

Paston near Mundesley – Stow Mill

Location: Stow Hill, Paston, NR28 9TG. West side of B1159, approx. 0.25 mile south of Mundesley and 3 miles north of Bacton. Privately owned and open to the public all year round.

This well-known four-storey corn mill in northeast Norfolk stands on the site of a medieval chapel and its name derives from the old English for resting place. Stow Mill is known to have existed since 1827, when it was conveyed to Thomas Gaze by his father James. It was handed down through the same family until William Gaze passed away

Stow Mill, Paston, April 2013.

in 1906. It was then bought by Mrs Mary Ann Harper, who leased it to her cousin, Thomas Livermore. He inherited the mill when Mrs Harper died in January 1928 and was the last miller at Paston before work ceased in 1930. While operational the whole mill was painted white, though the tower has been tarred for many years. The original common sails were later superseded by the more advanced patent type. As unlikely as it may seem, one pair of sails from the massive Upper Hellesdon Mill (see Chapter 5) were reused at Stow Mill, after being much reduced in size. A 5 hp steam engine was installed in the granary (built *c.* 1828 and converted to living accommodation in the early 1970s) around 1870 to provide additional power.

The redundant mill was purchased by Mr and Mrs Bell, who converted the tower into an annexe to the mill house and removed the machinery. The house itself was improved and extended. Douglas McDougal, famous for his successful flour business, was the owner between 1938 and 1954. In 1960 the mill and its associated buildings became the property of Mr C. M. Newton, who eleven years later conveyed them to his grandson, Mike Newton. Four new sails and a fantail were put in place in the early 1960s, and more work took place in the late 1970s and 1980s. This included fitting new steel stocks 52.25 feet (16 metres) in length and another set of skeleton sails. Repairs were made to the top of the tower and this allowed the polygonal dome-shaped cap to be turned for the first time in half a century. Some of the missing machinery was replaced, though the aim of returning the mill to full working order was abandoned on cost grounds. In the early years of the twenty-first century, further restoration has included repairing the cap, repainting the sails, stocks and tower, and replacing the fantail and fan-stage. Stow Mill was sold again in 1999 and was back on the property market in 2013. A plan to fully convert the mill to residential use in 2008 was opposed

Stow Mill, Paston, April 2013.

by S.P.A.B. Mills Section. Externally it is one of the most complete and impressive of Norfolk's remaining corn mills, despite being relatively small. It is highly photogenic and remains an attractive landmark on the coast road between Bacton and Mundesley.

WEST NORFOLK

West Walton – Ingleborough Mill near Wisbech

Location: Hill House Farm, Mill Road, Ingleborough, West Walton, PE14 7EU. 1 mile north of West Walton, approx. 5 miles north of Wisbech. Privately owned and not open to the public.

This is Norfolk's most westerly mill and it once vied with Terrington St Clement Balsam Fields Mill (see Chapter 5) as a contender for the county's tallest multi-sailer. Ingleborough Mill had an additional floor but whether this translated to an overall height advantage is unclear. It stands in countryside close to the borders with Cambridgeshire and Lincolnshire and has lost its sails and cap, but the eight-storey red-brick tower, complete with reefing stage on the fourth floor, still remains. It was advertised for sale as 'new' in June 1824, by the original owner Jarvis Palmer. The mill was built with five sails and gained a sixth sometime between 1846 and 1858. The tower contained four pairs of millstones and was topped with an ogee (onion-shaped) cap complete with a ball finial. By the mid-1890s a steam engine was providing additional power.

Ingleborough Mill, West Walton, 1928 (J. Neville).

Later owners included William Judd (*c.* 1863–92) and Henry Fretwell (*c.* 1896–1932), before the mill was purchased by George Garfoot in 1932 and became part of a modern milling operation. It was still working with all six sails in 1933 but by 1940 they had been taken down. The internal machinery was subsequently removed and a flat roof was fitted. Today the Grade-II listed building still retains its stairs and eight separate floors. Railings were fitted at the top of the tower in 1980. It is on private land and not open to the public. The mill and associated buildings were advertised for sale in 2012 and again in 2015.

Denver Mill near Downham Market

Location: Sluice Road, Denver, PE38 0EG. Off A10, 1 mile south of Downham Market. Site open to the public throughout the year but mill currently undergoing restoration.

Denver Mill is one of the best known in Norfolk and until recently was the county's tallest still in full working order. Situated in the southwest of the county on the edge of the Fens and a short distance from Denver Sluice, this is the last of twelve mills that operated within a 6-mile area. Despite several mills having more floors, this still ranks as one of Norfolk's tallest survivors being 59 feet (17.98 metres) high to the curb of the six-storey brick tower. The large ogee-shaped cap complete with gallery and topped with a finial, should add well in excess of 10 feet (3 metres) to the height. The tower is 26 feet 10 inches (8.17 metres) in diameter at its base and has a reefing stage on the third floor. It is rendered in a light brown, bordering on beige, colour, the end result (see photos) being much more pleasing than it sounds! The mill is visible for miles around and has long been a famous local landmark.

Denver Mill, 2004.

Denver Mill and its outbuildings, some of which are attached to the tower, were constructed in 1835–36 and replaced an earlier post mill. The first miller was John Porter, who worked it for around twenty years and died aged sixty-nine in 1858. The mill was taken over by a local farmer named John Gleaves, who passed it on to his son James in the early 1870s. He ceased milling in 1896. A steam mill had been added on the site by 1863 to supplement wind power. The steam engine was replaced with a Blackstone oil engine in the 1920s by Thomas Edward Harris, who had become miller around the turn of the century. His son, Thomas Edwin Harris, took over after his father's death in 1925 and continued to use the mill to grind animal feed until 1941. One sail had been struck by lightning two years earlier but Harris carried on milling until damage to the curb and other difficulties during the early years of the Second World War caused him to switch over entirely to the oil engine.

Following the death of Thomas Edwin Harris in 1969, his sister, Mrs Edith Staines, presented the mill to Norfolk County Council in the early 1970s. Extensive renovation took place during that decade including replacing the curb and repairing the cap and sails. Finally restored to full working order at the end of the 1970s, it was soon on the tourist map and open to the paying public all year round. Visitors were permitted to climb to the top of the tower, see flour being made, purchase bread, cakes and refreshments in the bakery and tea-room, browse the craft workshops and watch a video on the history of windmills. Unfortunately, due to insufficient wind, the sails didn't turn long enough during an enjoyable visit in 2004 for me to make my own camcorder video but I did witness them slowly revolving through the window while inside!

Denver Mill, 2006.

Between June 2008 and May 2013 the site was operated by the Abel family – trading as Denver Mills Ltd – who leased it from the owners, Norfolk Historic Buildings Trust. Work began in May 2009 to repair the sails and carry out other urgent repairs. The sails were back in position at the end of April 2010. Unfortunately, a major setback occurred on 4 October 2011, when part of a sail collapsed showering an area of the site with debris. A group of school children was visiting at the time but nobody was injured. Denver Mill was featured on *Alex Polizzi: The Fixer*, first broadcast on BBC Television on 6 March 2012, and *The Fixer Returns* in May 2013. The site is now under new management but the mill is lacking all its sails and both stocks for the first time in its long history. It was announced in October 2014 that funds were available to re-render the tower and carry out other essential work. A grant of £75,000 had been secured but the cost of bringing the mill back to full working order was estimated at around £750,000. It is likely to be years before this happens but it is hoped that eventually Denver will once again be the county's tallest working windmill.

Stoke Ferry Mill near Downham Market
Location: Boughton Road, Stoke Ferry, PE33 9ST. Off A134, approx. 7 miles southeast of Downham Market. Privately owned and available to rent as holiday accommodation.

Located in southwest Norfolk, Stoke Ferry Mill has lost its sails and cap and been converted to residential use, but despite an abortive demolition attempt the seven-storey brick tower has avoided the indignity of being truncated. It was constructed during the 1860s as a five-storey mill but two further floors were added around the turn of the century. This raised the height of the brickwork to approximately 58 feet (17.7 metres) and a new ogee (onion-shaped) cap measuring 10 feet (3.05 metres) tall and 17 feet (5.2 metres) in diameter was fitted. A set of four second-hand sails completed the transformation from medium- to large-sized mill.

Stoke Ferry Mill, September 2012.

Stoke Ferry Mill was still working in 1927 but by 1936 had lost its fantail and had only one pair of sails. In June of that year it was down to just one sail after the other crashed into the mill house causing damage to the building and injury to someone inside. According to local legend the mill survived efforts to demolish it during the 1950s using a traction engine but remained in a very sorry state until 1980. However, within two years it had been renovated and opened as Tower Mill Restaurant. Currently used for holiday accommodation, it is now attached to a large house and has a new roof with windows running round the top of the tower. Sadly, the ogee cap has been consigned to the dustbin of history. The tall and relatively narrow tower of Stoke Ferry Mill has some visual similarity with Moulton Mill in Lincolnshire before it gained its new cap, reefing stage and sails, but is just over 20 feet (6.1 metres) shorter to the curb of its tower.

Gayton Mill, near King's Lynn

Location: Litcham Road, Gayton, PE32 1PQ. On B1145, approx. 9 miles east of King's Lynn. Privately owned and currently part of a care home.

This mill was once eight storeys high (plus cap), which would have made it one of the tallest in the area. Located several miles to the east of King's Lynn, Gayton Mill still exists but lost its ogee-shaped cap and top floor before being converted to residential use. At this time it acquired a castellated top and a flat roof. Unusually for a tall mill, the tower is relatively narrow with interior and exterior diameters of 22 feet (6.70 metres) and 26 feet (7.92 metres) metres) respectively at base level. It used to have a reefing stage on the second floor and the tower still carries the scars. The mill was constructed in or just before 1824 and had several millers including Robert Matthews (1836–46) and Walter Hall (1853–72) before being purchased at auction

Gayton Mill, September 2012.

by Edward Lewis in 1873. He worked it for around two decades with the assistance of journeymen millers and it then stayed in the Lewis family for many more years. The Gayton Mills Co. was formed in 1912 and became a limited company in 1919. It stayed in business until 1937.

Originally fitted with four patent sails driving three pairs of stones and a six-bladed fantail, Gayton Mill probably stopped working by wind power around the end of the First World War and parted company with its sails around 1925, but retained its cap and fan stage intact until at least 1933. By 1872 an 8 hp steam engine was providing additional power and this was later replaced by paraffin and diesel engines. The mill continued to work by this means until 1937 but was derelict by 1949. It was sold in 1980 together with adjoining buildings, outhouses etc., and converted to residential use. It is currently utilised as part of a care home on the site. Gayton Mill stands close to the public highway and is a prominent local landmark.

Great Bircham Mill near Docking

Location: Snettisham Road, Great Bircham, PE31 6SJ. West of village approx. 2.5 miles south of Docking and 12 miles northeast of King's Lynn. Privately owned and in working order. Open to the public between Easter and the end of September only.

A well-known mill popular with visitors to Norfolk, Great Bircham Windmill was rescued from dereliction in the 1970s and painstakingly restored to full working order. Situated in the northwest of the county, it was built in 1846 by George Humphrey and the five-storey brick tower was later coated with tar to give the dark appearance

Great Bircham, Mill August 2015.

it retains to this day. The last miller before it stopped work in the 1920s was Billy Howard. The sails were taken down and the tower was left to deteriorate. For five years until 1888 the mill was worked by Joseph Wagg, whose great-grandson Roger and his wife Gina purchased the sad remains in September 1975. Together with millwright John Lawn, they embarked on the journey which eventually led to the sails turning again. A new ogee-shaped cap topped with a finial, four new sails and a fantail were fitted between 1979 and 1981, together with machinery recovered from other local mills. It was the first Norfolk windmill to be restored to full working order. Standing 52 feet (15.85 metres) tall to the curb of the tower, the total height is probably around 65 feet (19.81 metres) to the top of the finial. The diameter at ground level is 23 feet (7.01 metres) with 2 feet (0.61 metres) thick brickwork. The span of the sails is 67 feet (20.42 metres) and the cap weighs 8 tons.

In the twenty-first century, Great Bircham Windmill has been described by one writer as 'perhaps the most spectacular windmill in Norfolk' and by another as 'an archetypal mill in a perfect setting.' The rural location certainly adds to its 'chocolate box' charm, with the mill towering above its outbuildings and the surrounding countryside. Hungry visitors to Great Bircham can buy freshly made bread and take refreshments on board before climbing to the top of the mill. The cap is equipped with a gallery and it is possible to stand directly under the revolving fantail. There is also a reefing stage on the second floor and the original coal-fired ovens can still be viewed. While Denver Mill remains out of action this is currently Norfolk's tallest still in working order.

SOUTH NORFOLK

Wicklewood Mill, near Wymondham

Location: High Street, Wicklewood, NR18 9QA. Off B1135, approx. 2.5 miles west of Wymondham. Owned by Norfolk County Council and run by Norfolk Windmills Trust. Open to the public on selected dates.

Wicklewood Mill was constructed by Richard Mann during 1845–46. The first miller was John Browning Mallett, who worked at the site between 1846 and about 1875. He was superseded by William Palmer, who moved on in 1878 and was in turn replaced by James Doughty. He stayed till *c.* 1896. James Vout (1897–1902) and Walter Farrow (*c.* 1903–05) were later incumbents. Richard Mann sold the mill and mill house to Samuel Winter for £355 in 1882. The mill changed hands again in 1895 when Frederick Browes became its new owner. William Wade took over in 1906 after paying the princely sum of £330. He continued milling until shortly before his death in 1942 at the age of eighty-five, latterly in conjunction with his son Dennis. The mill stopped work at around this time.

The tarred five-storey tower is unusually slender, standing 43 feet (13.11 metres) tall to the curb with a compact internal base diameter of only 17 feet (5.18 metres). Four double-shuttered patent sails replaced the original common or cloth type at some point during the late nineteenth century and operated two pairs of stones. When it was donated to Norfolk County Council by William Wade's granddaughter Mrs Margaret Edwards during the 1970s, only a roofless shell was left. Norfolk Windmills Trust hired local millwrights Lennard & Lawn, who carried out a major restoration between the

Wicklewood Mill, September 2012.

Wicklewood Mill, after restoration April 2013.

winter of 1979 and the summer of 1981. This work included fitting a new boat-shaped cap, six-bladed fantail, 52-feet- (15.85 metres) long steel stocks, four new sails and much more besides. The total cost was estimated at around £12,500, which by today's standards seems a paltry sum to bring a ruined windmill back from the brink! The replacement cap was taken down for repair in 2005 and put back in place in July of the following year. It was decided to replace the steel stocks with traditional wooden ones. New sails were also fitted and the work was finally completed in October 2012. For some time prior to this the sails and stocks could be seen on the ground beside the tower. A 1920s Shanks paraffin engine that once provided supplementary power has been returned to Wicklewood after an absence of many years. Norfolk Windmills Trust hopes to return the mill to full working order in the near future.

Saham Toney, near Watton – Bristow's Tower Mill
Location: Ovington Road, Saham Toney, IP25 7HF. On east side of village, approx. 3 miles northwest of Watton. Privately owned and not open to the public.

The remains of Bristow's Tower Mill can still be seen in Saham Toney, though like many others it has been truncated as part of a conversion to residential use. In its working days it had six storeys and stood 50 feet (15.24 metres) tall to the top of the brickwork. The tower was topped by a boat-shaped cap and four patent sails. Built in 1826, it took its name from the family who ran it from new till the early 1900s. The first miller was John Bristow, who was superseded by his son (also named John) in 1845. Robert Bristow (son of John Bristow junior) took over in the early 1880s and stayed at the mill till around 1904. A date stone recording the year of construction carries the initials of John Bristow (senior) and his wife Sophia.

Bristow's Mill, Saham Toney, April 2013.

After it ceased working the mill stood empty before being converted in 1948. The cap and a pair of sails were still *in situ* in the mid-1920s but the machinery was later removed by Smithdales of Acle. However, the 8.5 feet (2.59 metres) great spur wheel was retained as a focal point of the new first floor lounge. After parting company with its cap and sixth floor the tower's exterior was painted white. A flat concrete roof was fitted and castellated brickwork was later added around the top edge. Since then the building has changed hands several times and various owners have carried out periodic repairs and alterations. Significant renovation work took place in 1960, and in 1980 a large single-storey extension was built around the tower. The mill was advertised for sale in 1983 and was back on the property market in spring 2013.

Caston Mill near Watton

Location: The Street, Caston, NR17 1DD. Off B1077, approx. 4 miles southeast of Watton and 8 miles west of Attleborough. Privately owned and not open to the public.

In addition to its other merits, this mill is worthy of note as the former home of the late millwright John Lawn. It has six storeys plus cap and measures 55 feet (16.76 metres) to the kerb of the brickwork. The cap would typically be expected to add about 10 feet (3.05 metres) to this figure. The total height to the tip of the topmost sail was said to be 102 feet (31.01 metres). The tower has a reefing stage on the second floor and the external diameter is 26 feet (7.92 metres) at the base and 17 feet (5.18 metres) at the kerb. The walls are 2.5 feet (0.76 metres) thick at ground level. In its heyday the mill had four patent sails and a six-bladed fantail. Three pairs of millstones were driven by wind power and a Ruston & Hornsby oil engine was added later to operate another pair.

Caston Mill April, 2013.

Caston Mill was built by local builder William Wright in 1864, with machinery installed by millwright Robert Hambling, for Edward Wyer, who continued running it until his death in 1897 at the age of seventy-six. His second son, James, took over the family business – which also included farming and baking – until his retirement in 1910. At this point it passed to James Wyer's brother-in-law Benjamin Knott, who worked the mill for thirty years. He was assisted in later years by his son Edward. It was during this period that the oil engine was installed to supplement wind power. The mill caught fire during a severe storm in March 1895. Originally a traditional corn mill, like many others it switched to grinding animal feed towards the end of its working life. After being sold to James Bilham in 1940, Caston Mill stopped working by wind power. John Lawn purchased the mill and former granary attached to it as his private residence in 1969. He removed the cap and sails for repair in November 1983, and they were back in place by 1985. Mr Lawn's professional commitments maintaining and renovating many other local mills prevented him from fully restoring Caston Mill before his untimely death. Today the building is still privately owned and is located at the end of a long driveway. It is visible from the road and retains its cap and part of the fantail but has no sails.

Great Ellingham Mill, near Attleborough

Location: Long Street, Great Ellingham, NR17 1LL. Approx. 3 miles northwest of Attleborough, off A11. Privately owned and not open to the public.

Until recently a derelict tower but now having been renovated and converted to residential use, Great Ellingham Mill is 53 feet (16.15 metres) tall to the top of the brickwork and the walls are 2 feet (0.61 metres) thick at the base. The interior diameter

Great Ellingham Mill, April 2013.

at ground level is said to be just 18 feet (5.48 metres). The mill was built by Jeremiah Fielding around 1849 with five storeys and later raised to its current height with the addition of a top floor. Double-shuttered patent sails replaced the old common cloth type after the tower was heightened. Two pairs of overdriven stones were installed on the first floor, which reputably had a 15 feet (4.57 metres) high ceiling. This would explain the large gap between the windows on levels one and two.

Many millers worked at Great Ellingham over the years including Jeremiah Fielding himself from 1850 to 1854, though James Buck (1849–1850) was probably the first. Samuel Le Grice purchased the mill and associated buildings in 1854 and remained the owner till shortly before his death in 1905. Among the journeymen millers he employed was William Stackwood, who was listed as miller in 1861 and sadly experienced far more than his fair share of personal tragedy. In addition to the deaths of his two young daughters and then his wife between 1863 and 1866, the unfortunate Mr Stackwood also became bankrupt in June 1866. George Butler was the longest serving miller at Great Ellingham, clocking up thirty-one years starting from 1865. The nearby mill house and bakery were destroyed by fire around 1900 but the mill itself was spared. It continued to work by wind power until at least 1916 before later being stripped of its sails, fantail and cap. The sails and stocks were reused on another mill. Milling continued at Great Ellingham using an oil engine until the mid-1920s, when the mill closed. It reopened as part of a bakery business just after the Second World War and was still in use into the 1970s, before gradually falling into a state of disrepair. Happily, the old building is currently enjoying a new lease of life after permission was granted to convert it to living accommodation and build a large wooden extension adjoining the tower. Work began in 2009 but did not include replacing the cap.

Old Buckenham Mill near Attleborough

Location: Mill Road, Old Buckenham, NR17 1SG. Off B1077, 5 miles southeast of Attleborough. Owned by Norfolk County Council and open to the public on selected dates between April and September. Currently awaiting new sails.

Though hardly a giant in terms of its height, the windmill at Old Buckenham is nevertheless a record-breaker in several other aspects of its construction. For starters, its boat-shaped cap measures 24 feet (7.32 metres) in width making it the largest in the country and the whole cap assembly weighs around 14 tons. Add to that the widest sails of any surviving British mill at 10 feet 4 inches (3.15 metres), the most pairs of millstones (five) on one floor and what is thought to be the biggest great spur wheel at 13 feet (3.96 metres) in diameter! To accommodate the huge cap it follows that the red-brick tower's diameter at the curb is the largest in the UK at 23 feet (7 metres), though its external diameter of 26 feet 6 inches (8 metres) at ground level is beaten by many mills. The sails have been variously reported as between 75 feet (22.86 metres) and 79 feet (24.07 metres) in diameter. The walls are 2 feet (0.6 metres) thick and the height of the five-storey tower is just 42 feet (12.8 metres) to the curb. The overall dimension including the cap is 54.5 feet (16.61 metres), producing an unusually squat appearance as if it was originally intended to be much taller. There appears to be no firm evidence to support this, though the original owner is said to have been forced to utilise cheaper materials in the upper stages due to cash flow problems.

Built in 1818 for John Burlington (who died in 1853 aged seventy-nine) on the site of earlier post mills and owned from 1862 to 1900 by famous mustard makers J. & J. Colman, Old Buckenham Mill was used to grind corn for flour and animal feeds. Prince Frederick Victor Duleep Singh was another high-profile owner between April 1900 and March 1905. It was originally equipped with no less than eight fabric covered

Old Buckenham Mill, March 2009.

sails – making it one of very few eight-sailers in Norfolk – but these were replaced by four new ones following severe storm damage in 1879. A granary built close by in 1856 once housed a 12 hp steam engine but the building was later converted to living accommodation and is still in existence. The mill finally stopped working in 1926 after losing its fantail. The last miller was William Goodrum, who had gamely carried on working the mill despite having an arm amputated in 1921 following a horrific accident involving the machinery.

After it stopped working the mill gradually fell into a state of disrepair and what remained of the cap and sails were removed in 1976 by well-known local millwright John Lawn, who later carried out an extensive restoration between 1992–96. A new cap and four sails were fitted to complete this process in 1996. On an enjoyable visit to Old Buckenham Windmill in 2008, I was informed that the original intention was to return it to full working order but the deteriorating health of Mr Lawn, who sadly died from cancer in late 1999 at the age of sixty-three, meant that this aim became unrealistic. Nevertheless, the mill was in excellent condition and the sails were capable of turning,

Old Buckenham Mill and former granary, March 2009.

but it was unable to grind corn as first hoped. A memorial plaque to the late millwright is situated at the entrance of the mill. Now owned by Norfolk County Council, it opened to the public in 1997 and is available for viewing on open days. It is well worth the effort to climb into the cap and experience its massive dimensions at close quarters from the inside, though the external gallery is out of bounds. Unfortunately the sails were removed due to rot in 2010 and it was hoped that around £32,000 could be raised to pay for the manufacture and installation of a new set. The mill was closed to the public for about two years before reopening on selected dates from June 2011. It is still possible that, funds permitting, Old Buckenham Mill may one day be fully restored to working order.

Billingford Tower Mill near Scole

Location: Billingford Common, IP21 4HL. Just off A143, approx. 1 mile east of Scole and 3 miles east of Diss. Owned by Norfolk County Council and run by Norfolk Windmills Trust. Open to the public on selected dates. Currently due to be restored to working order.

Being the first windmill I ever climbed as a child, this impressive example retains a special place in my affections. At the time it seemed enormous but in reality it is a medium-sized corn mill with a five-storey red-brick tower. Billingford Tower Mill stands in splendid isolation on a common close to the A143 in South Norfolk near to the border with Suffolk, though at various times it has shared its surroundings with a variety of other buildings that have since been demolished. It was constructed in 1860 on the site of Billingford Post Mill dating from around 1797, which was destroyed during a gale on 22 September 1859. The miller, George Goddard, was inside at the time but escaped with minor injuries while another man (said to be either an employee or a customer) was killed. Mr Goddard then became the first tenant of the new mill, built for William Chaplyn by W. Skinner for the sum of £1,300.

Billingford Mill, October 2012.

The present tower mill has the distinction of being the last in Norfolk to work commercially by wind power and also the county's most southerly mill. Apart from a break during the Second World War, she worked continuously for the best part of a century and saw at least thirteen different millers come and go. One admitted in 1872 that he had added starch fibre to barley meal supplied to a customer! Another, Edward Woodrow left to take charge of Scole Tower Mill (no longer standing) and later moved on to Horsford Mill (now partly converted to residential use). His son Charles, who eventually bought and worked the eight-storey Peafield Mill at Lakenham, described Billingford as 'one of the prettiest mills ever built'. Following the death of William Chaplyn in 1881, the mill failed to sell at auction and continued to be leased.

The last commercial miller at Billingford Mill was Arthur Daines, whose father George had bought it in 1924. Arthur joined his father's business in 1933 before serving in the Royal Navy during the Second World War. He returned to Billingford and continued milling after carrying out repairs. Having been reduced to just one pair of sails, further problems were encountered and the mill ceased operating by this means in 1956. Arthur Daines managed to keep the mill open for another three years using an oil engine that had supplied supplementary power since replacing an earlier steam engine. The mill was purchased by Victor Valiant in 1959 and he later donated it to Norfolk County Council following extensive restoration during the early 1960s. The work was undertaken by Thompson's of Alford and included the replacement of the sails and a stock. The total cost was nearly £4,000, partly paid for by grants and local fund raising. Around this time the buildings near the mill were torn down and the black tarred cap was painted white. The maintenance of the mill since then has been the responsibility of Norfolk Windmills Trust. A new fantail was fitted in 1976 to replace one badly damaged in a gale. In July of the following year, millwrights Lennard & Lawn installed a 52 feet (15.85 metres) long fabricated steel stock in place of a wooden one fitted in 1962.

Billingford Mill was eventually returned to full working order in 1998. A team of volunteers operated it on a part-time basis during open days and a lady named Linda Joslin became the first female miller on the site in 2002. It was later reported that she left after a long-running dispute over the maintenance of the building. On a visit during an open day in 2007 the sails were turning merrily, though they were removed due to rot in June 2009, leaving just the stocks. It was announced in early 2014 that work to replace the sails and stocks and restore the cap was expected to commence in April 2015, with the aim of once again returning Billingford Mill to full working order.

CENTRAL NORFOLK

East Dereham Mill
Location: Greenfields Road, off Norwich Road, East Dereham, NR20 3TE. From A47 take slip road marked Swanton Morley and follow signs to mill. Run by Dereham Town Council and open to the public on Wednesday, Saturday and Sunday. Exterior is accessible at all times.

The recent history of this mill has – like several others – been a story of extensive renovation followed by a period of gradual decline, before funds eventually became

East Dereham Mill after restoration, October
2013.

available to enable further repairs to take place. After being purchased for the nominal sum
of £1 by Breckland District Council from Greens (Nurseries) Ltd in 1978, restoration took
place in 1986–87 including the installation of a new cap and sails. This was undertaken by
John Lawn, who had also been involved in removing the remains of the previous cap and
windshaft in 1977, in conjunction with the Manpower Services Commission Community
Programme. The work was said to have cost £30,000 and the mill was opened to the public
on 14 September 1987. It became the responsibility of Dereham Town Council fifteen years
later. Part of a sail was blown off in January 2004, and all the remaining sails were taken
down in August 2006, leaving just the pair of stocks. After an application for a substantial
lottery grant was turned down, limited funds allowed only essential maintenance to be
carried out and the building was closed to the public. Major restoration reportedly costing
£75,000 finally took place over several months during 2013. This culminated in new sails
being fitted to the existing steel stocks and the mill officially reopened as a community
exhibition centre on 7 September 2013. Though not in working order, in appearance it is
now one of the county's finest examples.

   East Dereham Mill was built in 1836 by millwright James Hardy and miller/baker
Michael Hardy. It sold at auction for £650 eight years later, becoming the property of
William Fendick. Following his death at the age of forty-eight in 1861, the mill stayed in
the same family until the early twentieth century. It became known as Fendick's Mill and
also as Norwich Road Mill. Charles Gray was reported to have paid just £450 for the
property in 1909. To the curb of the five-storey tower it stands 42 feet (12.8 metres) tall,
with internal and external base diameters of 20 feet (6.1 metres) and 25 feet (7.62 metres)
respectively. A boat-shaped cap complete with iron-railed gallery topped the tower, and
the fantail was a six-bladed type. The four sails drove three pairs of stones and by the
early 1860s a steam engine was providing auxiliary power. This was later superseded
by a diesel engine which also operated an additional pair of stones. The mill switched
over to producing animal feed during the First World War and the sails were taken
down around the time of Charles Gray's death in 1922. The cap and windshaft were

left in place and the mill continued to work by engine power until 1937. An application for permission to convert the tower to residential use after the Second World War was turned down. After falling into a state of disrepair, it was purchased by Reginald Green of Green's Nurseries and in 1973 survived the threat of demolition having earlier gained Grade II listed building status. The mill stands beside a track known as Cherry Lane off Norwich Road, though motorists should take Greenfields Road and continue to the far end where the mill is located on a green close to a large housing estate.

Yaxham Tower Mill near East Dereham

Location: Norwich Road, Yaxham, NR19 1RH. On B1135 off A47, approx. 3 miles southeast of East Dereham and 15 miles west of Norwich. Privately owned and run as part of a business.

Yaxham Mill was constructed in 1860 by William Critoph and the six-storey tower stood 48 feet (14.63 metres) tall to the curb, plus an ogee cap topped with a ball finial. The interior diameter at the base was 20 feet (6.1 metres) and a stage was fitted at second floor level. The mill, which survived a lightning strike in 1872, was equipped with four double-shuttered sails and three pairs of stones. A bakery and shop stood nearby. Mr Critoph also had a post mill and a smock mill on the same site but the latter was probably demolished soon after the new tower mill became operational. The post mill survived until about 1915 and was purchased along with the tower mill in 1904 by Thomas William Stebbing Parlett for the total sum of £780. A steam engine – later superseded by diesel power – provided additional urge from around 1880. After the sails were taken down in 1922 the tower mill slowly deteriorated but still retained its cap and windshaft until at least the late 1930s. The machinery was removed in 1940.

In recent years Yaxham Mill has changed hands several times and was given a new lease of life as part of a restaurant and B&B business on the site. During the 1990s the mill and its outbuildings were extensively renovated and the tower still has all six floors intact, but a conical roof has replaced the original cap.

Yaxham Mill, September 2013.

Chapter 2

Tower Windpumps (Drainage Mills)

There is some controversy over whether windpumps should be called mills as they lack millstones and were built to lift water from marsh land instead of grinding corn. Technically they are different machines but to the casual observer they look virtually identical. An alternative name of drainage mill is commonly used in lieu of windpump and modern authors and websites often refer to them as mills. For this reason I will continue to use that term, even though purists may disagree! Virtually all of Norfolk's surviving tower windpumps are located in the east of the county, particularly around the Broads. I have grouped them according to which river they are associated with.

RIVER YARE

Reedham – Berney Arms Mill
Location: Reedham, NR30 1SB. North bank of River Yare, west of Breydon Water, 3 miles northeast of Reedham. Owned by English Heritage and open to the public on selected dates. Accessible by river and rail but inaccessible by road.

Standing 70 feet 6 inches (21.49 metres) to the top of its white boat-shaped cap, Berney Arms Mill is by far the tallest drainage mill or windpump on the Norfolk Broads and possibly also Britain's tallest. The internal diameter of the seven-storey, tarred-brick tower is 28 feet (8.53 metres) at its base. It has a white-painted iron reefing stage on the third floor and a gallery on the cap. According to Rex Wailes, a leading twentieth century authority on windmills, 'the span of the sails is almost the same as the height of the mill, and it will be noted that they are not identical, two having come from another mill.' However, it has also been claimed that the sails are the longest of any surviving British mill at 85 feet (25.9 metres) in diameter. This is 12 feet (3.66 metres) more than the sail span of Sutton Mill (see Chapter 1) and roughly the same as that of Southtown (Cobholm) High Mill (see Chapter 5), which was over 40 feet (12.19 metres) taller. Taking the height of the tower and the length of the sails together, Berney Arms Mill cannot be far behind Sutton Mill in total height. Although Berney Arms has only seven floors compared to Sutton's nine, the difference in the heights of their respective towers (caps included) is less than 10 feet (3.05 metres).

Berney Arms Mill, Reedham, August 2009.

Built for the Berney family, who owned much of the land around Reedham for hundreds of years, the exact date of its construction is uncertain. The earliest date pencilled inside the cap is 1870 but she is generally considered to have been built around 1865 by the Great Yarmouth millwright Edward Stolworthy, on the site of earlier mills. A drainage mill stood close to the present site as early as 1797 and by 1822, a cement works complete with windmill and steam engine was in existence. A five-storey mill and 12 hp steam engine were mentioned in the *Norfolk Chronicle* in 1847, while the same publication thirteen years later spoke of 'a 24 hp brick tower patent sail windmill'. The present seven-storey high mill's well-proportioned tower shows no signs of having been hained (heightened) and it is generally considered that it was completely new from the ground up rather than a rebuild of the previous one. It was probably unique in being used originally to grind cement clinker. Cement was produced onsite from chalky mud dredged from the river bed, chalk delivered by wherries from Norwich and Whitlingham, and also from clinker brought over from the steam-powered Burgh Castle Cement and Brickworks located a short distance across the water.

Following the closure of the cement works at Berney Arms in 1880, within three years the mill had been converted to draining the marshes, probably by the millwright Richard Barnes. This necessitated the construction of a massive scoop wheel housed in a weather-boarded structure known as a hoodway, situated a short distance from the mill. The scoop wheel, which still exists, measures 24 feet (7.32 metres) in diameter, making it one of the largest of any drainage mill. It is said to have revolved seven times for every twenty turns of the huge sails, raising water to a height of 8 feet (2.44 metres). Standing on the Berney Marshes at the head of Tile Kiln Reach on the River Yare, close to Breydon Water, the mill is around 3 miles from the village of Reedham and around 5 miles from the holiday resort of Great Yarmouth. Accessible by river, rail and 'shank's pony', it is miles from any proper road but is a well-known landmark standing out against the typically flat Norfolk landscape. It was recently featured in the opening credits of the ITV television series *Kingdom*, with the enormous offshore wind turbines on Scroby Sands in the background. Between 1883 and 1949, when it was superseded by an electric pump, Berney Arms Mill was worked by a succession of marshman-farmers living with their families at nearby Ash Tree Farmhouse. The mill was taken over by the Ministry of Public Buildings and Works in 1951 and opened to the public in May 1956. Following damage to the sails, fantail and gallery in the winter of 1961–62, a full restoration took place in 1965–67. More work was carried out in 1972–73, including repainting and retarring. An exhibition of windmills of eastern England was created inside the tower and still exists today, with pictures of local mills and text by Rex Wailes. The RSPB later took over much of the land in the locality and the Berney Marshes Reserve was opened in 1986. Ash Tree Farmhouse underwent major renovation work in 1989 and was converted into flats for RSPB staff.

Now owned by English Heritage and having the dubious distinction of being their most remote and inaccessible building, the mill underwent major restoration between 1999 and 2007 as part of the Land of the Windmills project, which also saw the repair of a number of smaller mills in the area. During this period she was sail-less and a shadow of her former self. The work was originally expected to take four years to complete and cost in the region of £350,000. The cap was removed by crane on 4 November 2002, and refitted after repair on 6 May 2003, but the fantail was not replaced until 22 April 2007. The sails were scheduled to be refitted in September 2003, but it wasn't until 23 May 2007, that the first pair was put back in place. The second pair finally followed on 2 June 2007. The sails had reportedly been on a round trip of over 200 miles to and from a millwright near Henley-on-Thames, while the cap had travelled to Leicestershire and back. Although once more restored to almost her former glory, the mill is not in full working order. It was finally reopened to the public on selected days in the summer of 2009, a full decade after closing its door. Boat trips from Great Yarmouth – complete with a guided tour of the mill – restarted in August of that year. My belated return visit was an enjoyable experience, though the reefing gallery on the outside of the tower that I walked round as a youth is no longer accessible to the public. Berney Arms Mill is definitely one of my favourite windmills. The stark but picturesque (and allegedly haunted) location, tall black-tarred tower and contrasting massive white-painted sails, combine to make it one of the county's most photogenic mills. From a personal perspective a distant family connection also makes it special. William Hewitt, well-known locally as 'King Billy', was a marshman-farmer at Ash Tree Farm from around 1885 to 1924. He married my great-great grandmother Susannah Hurrell, after they had both been widowed (see photo). William's son James and his wife Eunice took over the running of the farm including the windmill between 1924 and 1946. They were assisted in later years by their son Reggie and his wife Nora, who also lived there until 1946, when Henry Hewitt took over. Known locally as 'Yoiton', he was a grandson of 'King Billy' and was the last to run the mill before it stopped working. His granddaughter Sheila Hutchinson has done much to put Berney

Top: William and Susannah Hewitt, Berney Arms *c.* 1900. *Bottom*: Berney Arms Mill from the River Yare August 2009.

Arms and Reedham on the map in recent years with a number of books about the area and the people who lived there.

Reedham Marshes – Polkey's Mill and Cadge's Mill

Location: Seven Mile House, Reedham, NR13 3UB. North bank of River Yare, 2 miles northeast of Reedham. Owned by Norfolk County Council and open to the public on selected dates only. Accessible by river but inaccessible by road.

Several other drainage mills still exist on the Reedham and Halvergate marshes, in various states of repair. Two recently restored examples are located about a mile to the south of Berney Arms. Polkey's Mill was returned to full working order between 2002 and 2005 by millwright Vincent Pargeter for the Norfolk Windmills Trust and despite only being three storeys high carries sails with a relatively large diameter of 70 feet (21.34 metres). The sails differ from those of most mills in that they turn in a clockwise direction. Vincent Pargeter fitted the new sails in September 2005, replacing what was left of the previous ones taken down in March 2003. A windpump existed on the site in the late eighteenth century, though the present one is thought to date from sometime after 1840. The unusual name probably derives from a nickname given to a Mr Thaxter – a marshman who operated the mill over quite a long period – but at various times it has been called South Mill or Seven Mile Mill. The tower was definitely increased in height at least once and probably twice during its working life, which ended in the early 1940s. The sails drove a scoop wheel to pump water from the marshes and from 1880 were assisted by a steam engine made by Richard Barnes. This no longer exists but the engine house beside the mill has recently been restored.

Cadge's Mill is situated a short distance from Polkey's Mill and like its close neighbour can be seen from the railway line that connects Reedham and Great Yarmouth. It has also been known by various other names including Kedge's, Stimpson's and

Left: Polkey's Mill 1899. *Right*: Cadge's Mill *c.* 1930s, Reedham. (P. Allard)

Batchie's Mill. According to different sources the mill's scoop wheel measured between 16–20 feet (4.88–6.1 metres) in diameter and was unusual in being enclosed in the tower rather than outside. As a consequence the tower itself is wider than normal for a windpump and also incorporated a fireplace. It now houses switch gear for the electric pump which eventually replaced the diesel one that made the mill redundant! After it ceased operating in 1941 the sails were taken down by Reggie Hewitt and recycled for use elsewhere. Most of the internal machinery was later removed. Like Polkey's and Berney Arms, Cadge's Mill was restored during the early twenty-first century as part of the Land of the Windmills project. It has a new boat-shaped cap fitted in 2006 but currently lacks sails and fantail. It is hoped that these may be installed at some point in the future.

RIVER BURE

Horning – St Benet's Abbey Mill
Location: NR29 5NU (Hall Road, Ludham). North bank of River Bure, approx. 1.25 miles west of Thurne Mouth. Accessible by river, on foot or by car. From Ludham Bridge heading towards Ludham on A1062: turn right onto Hall Road, then take track signposted to ruins. Parking available onsite. Owned by Norfolk Archaeological Trust and open to the public at all times.

As far as unusual locations go, few would expect to find the remains of a windmill built into the stone gatehouse of a ruined abbey miles from anywhere. In this respect it is almost certainly unique and may have been the first brick tower mill to be erected in Norfolk. The remains are certainly the oldest of their type left in the county but are relatively recent compared with the surrounding stonework. St Benet's Abbey began to flourish in the eleventh century and grew in size and power over the next few hundred years, though its humble origins can be traced back to the early ninth century. It was the sole survivor in Britain of the Dissolution but by 1545 was deserted and much of the stone was later used at other sites.

St Benet's Abbey Mill is thought to have been built sometime between 1728 and 1740 but an earlier mill existed elsewhere on the site at the beginning of the eighteenth century. The gatehouse was the only part of the ruined abbey left relatively intact, though it was necessary to demolish the top section to accommodate the sails! A platform was constructed around the top of the truncated gatehouse to serve as a reefing stage. The cap was also noteworthy, being conical shaped with horizontal boarding. As no fantail was fitted the cap had to be turned manually by the miller according to the wind direction. The four original sails were the rather primitive common type, though later two common and two patent sails were used. The mill's original role was probably to crush cole or colza seeds to produce oil for lamps. It may have been converted to a drainage mill by 1810 and perhaps even as early as 1781, with the addition of a scoop wheel. It is unlikely that it worked again after the cap and sails were blown off in 1863. The last millman was William Grapes, who also ran a public house nearby.

St Benet's Abbey Mill and gatehouse ruins. (*Top*: May 2009; *bottom*: March 2011)

After it stopped working the mill soon became a roofless, derelict shell, and over the years much of the interior wall became covered in graffiti left by visitors. At some point the scoop wheel and the platform on top of the gatehouse were removed. Renovation work was carried out on the tower and gatehouse in 1929. The site was surrounded by scaffolding in the summer of 2012 to facilitate the latest round of repairs and restoration. The mill and abbey ruins are officially in Horning but are closer to the village of Ludham. The mill tower can be seen at a distance from Ludham Bridge and the site is accessible by water or by means of a long narrow track. Walkers face quite a hike but a small car park is provided close to the gatehouse. Some say that the ruins are haunted and have a strange atmosphere even in daylight. Having visited on several occasions they seem to exude a slightly other-worldly quality, and perhaps have a weird kind of stark beauty. According to local legend a monk who turned traitor was hanged from a beam over the gatehouse, and the horrific episode is allegedly replayed in sound and vision on the night of 25 May each year.

Upton Black Mill, near Acle

Location: Upton, NR13. West bank of the River Bure north of Upton Dyke and just northwest of Oby Mill. Privately owned and not open to the public. Not accessible by road but the exterior can be viewed at all times from a footpath running alongside the river. Car parking is available at Upton Dyke (NR13), 1 mile northeast of Upton village (follow signs to Boat Dyke), near Acle.

The date 1800 is recorded in white on the tarred tower of Upton Black Mill. On some Norfolk Mills (see Hunsett and High's Mill, Potter Heigham, for example) the date shown cannot be taken as an accurate indicator of the year of build, but in this case it is verified by documentation. The mill was elevated to its present height

Upton Black Mill, July 2013.

about a century later and is sometimes called Upton Tall Mill. Five iron bars encircle the tower, which is topped by a white boat-shaped cap with gallery. Though the fan stage remains it no longer houses a fantail and the sails were removed many decades ago. The mill keeper's cottage was being hired to holiday makers by the early 1970s and the mill itself was said to have been converted to the same use by the mid-1980s. The engine house still contains a Ruston and Hornsby diesel engine, which replaced an earlier steam engine.

Upton Black Mill is a half-hour walk from Upton Dyke car park (see Palmer's Mill, Chapter 4) along the west bank of the River Bure. There was also once a corn tower mill near Upton village green but this was pulled down around 1890.

OBY MILL

Location: Oby, NR29. East bank of the River Bure north of Upton Dyke and just southeast of Upton Black Mill. Privately owned and not open to the public. Not accessible by road but can be seen from footpath on the west bank of the river. Car parking is available at Upton Dyke (NR13).

The tower drainage mill at Oby is thought to have been constructed as early as 1753 and is almost certainly the oldest surviving windpump on the Broads designed specifically for this purpose. The remains of St Benet's Abbey Mill are older but it was most likely built to crush seeds and converted to drainage at an unknown date. Oby Mill had four floors and was equipped with four patent sails which drove a scoop wheel and also a saw bench. It appears to have been one of the longest working before the sails were dismantled in 1933. However, it may not have operated by wind power

Oby Mill, July 2013.

up to this date as steam and diesel engines were also in use at various stages. The latter continued to drive a turbine for a while after the removal of the sails. The mill stands in the combined parishes of Ashby with Oby and was worked for much of its life by members of the Wiseman and Davey families. Mr Herbert Davey was in charge in 1933. By the 1980s Oby Mill was derelict and without a cap or sails, though the remnants of the stocks and fan stage were still in position. These were taken down at the end of the twentieth century and a temporary cover was fitted to the top of the tower to help prevent further deterioration. At the time of writing this is still in place. The windshaft (to which the sails were connected) can still be seen.

It was reported in April 2008 that Adam Whiting paid £41,500 for Oby Mill and its outbuildings at auction, apparently outbidding Norfolk Windmills Trust who were also interested in acquiring the Grade II-listed property. It was advertised for sale again in March 2014 and now has new owners. Long-term plans may involve fitting a new cap and sails and possibly even converting the mill to produce electricity. It can be clearly seen across the water from a footpath leading from Upton Dyke car park along the west bank of the River Bure.

CLIPPESBY MILL

Location: Clippesby, NR29. East bank of the River Bure, almost opposite opening to Upton Dyke and south of Oby Mill. Privately owned and not open to the public. Not accessible by road but can be seen from footpath on the west bank of the river. Car parking is available at Upton Dyke (NR13).

Clippesby Mill, July 2013.

Standing in the combined parishes of Ashby with Clippesby this is another old mill possibly dating back to the late eighteenth century. A date of 1814 was found inside but this is thought to refer to an extra floor being added rather than the year of build. The later brickwork at the top of the tower contrasts with the narrower original bricks. The four-storey tower stands 41 feet (12.5 metres) tall including the boat-shaped cap and its base diameter is 21 feet (6.40 metres). The windshaft can still be seen but the sails and fantail disappeared long ago. The sail stocks were taken down after a lightning strike in 1978. A steam engine and turbine once supplied supplementary power and the engine house still exists. Some internal machinery is still *in situ* in the tower but the external scoop wheel has been removed. Although looking virtually untouched from the outside, Clippesby Mill was converted to residential accommodation in the late 1950s but even during its working life was apparently inhabited. There is evidence to suggest that it once had a staircase between the ground and first floor levels and a sleeping area. A fireplace on the ground floor has since been bricked up. Now Grade II listed the mill is owned by millwright Vincent Pargeter. It can be viewed from the west bank of the River Bure at Upton Dyke. A small restored hollow post windpump known as Palmer's Mill (see Chapter 4) can also be seen at Upton Dyke.

Tunstall – Stracey Arms Mill, near Acle

Location: Tunstall, beside A47 between Acle and Great Yarmouth, NR30. Limited parking onsite. Also accessible from the south bank of the River Bure. Owned by Norfolk County Council and open to the public daily from Easter until the end of September.

Standing close to the very busy A47 Acle New Road, the Stracey Arms Windpump is arguably seen by more people every day than any other Norfolk mill. It is sandwiched between the road and the River Bure, where holiday boats proliferate during the summer months and can also be seen from the Acle to Great Yarmouth railway line. The drainage mill we see today was built in 1883 by the Great Yarmouth millwright Robert Barnes. It was necessary to sink piles and a pine raft into the ground to provide a firm base on which to build. Originally equipped with a scoop wheel, it later switched to a turbine pump. An earlier windpump on the site had been worked by members of the Arnup family since 1831 and they took over the present mill from new. The same family continued to run it until the 1940s and during this period it was also well known as Arnup's Mill. For many years a nearby public house named the Stracey Arms was a popular watering hole between Acle and Great Yarmouth. The last millman before the sails drew to a final halt in 1946 was Fred Mutton, who had taken charge a few years earlier. The mill also served as a fortified post during the Second World War, which resulted in some damage to the tower. Despite this, the mill could have continued working for longer but the decision was made to install a 20 hp electric pump.

By the time restoration began in late 1960, it was necessary to repair the cap and brickwork and to replace the fantail, stocks and sails. The mill was signed over to Norfolk County Council by its owner, Lady Stracey, and later opened to the public.

Stracey Arms Mill, Tunstall, June 2010.

For many years it was in excellent non-working condition with skeleton sails, but in the early twenty-first century it became apparent that more work was required. It can be seen from the photograph that the topmost sail was absent in the summer of 2010. At the time of writing all sails have been removed, though the stocks remain. Part of the fantail is still in place but all the blades are missing. It is hoped that repair work can commence as soon as funds allow.

Mautby Mill, *near Great Yarmouth*

Location: Mautby, NR29 3JD. North bank of River Bure, approx. 1 mile west of Ashtree Farm Mill and 3 miles northwest of Great Yarmouth. Privately owned and not open to the public. Not accessible by road.

Situated on the River Bure between Stracey Arms and Ashtree Farm mills, this restored windpump has a house attached to its tower and was converted into a residence in the early 1980s. The work included the installation of a new cap and sails to replace the decayed originals. Since then it has changed hands on at least two occasions and stands on private property some distance from the village of Mautby. One of the sails was damaged during a gale in 2007. The present mill was preceded by a skeleton windpump dating back to the eighteenth century. A Ruston and Hornsby oil engine supplemented wind power in 1919 and later took over as the sole power source for many years. By the early 1970s the mill was derelict before being rescued a decade later.

Mautby Mill, June 2010.

Ashtree Farm Mill, near Great Yarmouth

Location: Ashtree Farm, south bank of River Bure, approx. 1 mile east of Mautby Mill and 2 miles northwest of Great Yarmouth. Privately owned and not generally open to the public. Access to interior by prior arrangement with Norfolk Windmills Trust. Not accessible by road.

Standing alone beside the River Bure some distance from the A47, this picturesque little windpump could genuinely claim to be in the middle of Nowhere. Bizarrely, that was the official name bestowed on the area in the Assessment Act of 1862. Like Horsey Mill, this recently restored example at Ashtree Farm was one of the last brick tower drainage mills to be built in Norfolk. It was constructed in 1911/12 on the site of an earlier windpump by Smithdales of Acle for the Ecclesiastical Commissioners and continued to work until January 1953, when it was severely gale damaged. By the 1980s the mill was derelict with no cap or sails but retained its internal machinery. Local millwright Richard Seago carried out a major restoration over several years in the early twenty-first century, which culminated in the installation of a new cap, fantail, windshaft and skeleton sails, in 2006.

Ashtree Farm Mill is 26 feet (7.92 metres) tall to the top of the tarred three-storey brick tower and 34 feet (10.36 metres) including the boat-shaped cap. The base diameter is 14 feet (4.27 metres) and the external scoop wheel measures 16 feet (4.88 metres) in diameter. Now leased to the Norfolk Windmills Trust the interior can be viewed by appointment only. The exterior can be seen at close quarters from the

Ashtree Farm Mill, June 2010.

river or at a distance from the Acle Straight (A47) just outside of Great Yarmouth. From the road the difference in size and design compared with the huge modern offshore wind turbines on Scroby Sands is remarkable – creating a striking contrast between ancient and modern while retaining a semblance of continuity.

RIVER THURNE

Horsey Mill, near Sea Palling
Location: Horsey Staithe, NR29 4EF. Beside the B1159 between Waxham and Somerton. Also accessible by river: at east end of Horsey Mere off the River Thurne, approx. 2.5 miles northeast of Potter Heigham. Owned by the National Trust and open to the public between March and the end of October.

Having been completed as late as 1912, this four-storey drainage mill was one of the last of Norfolk's tower windpumps to be built, though it stands on the foundations of earlier versions. The Horsey Black Mill possibly dated back to the eighteenth century but it seems that there were two rebuilds on the present site during the following century, the last being in 1897. The present mill was constructed on the same base and has one of the widest towers of any remaining Broads windpump. The diameter of the sails is about 65 feet (20 metres). Being one of the county's best known and most photographed mills it is popular with visitors in the summer months and is open to the public from spring through to autumn. It stands at Horsey Staithe just by the Mere and is a prominent landmark on the old coast road. Built by the millwrights England's of Ludham, it utilised a turbine pump rather than a scoop wheel to drain water from the

Horsey Mill, September 2012.

surrounding land. A steam engine supplied auxiliary power before a diesel pump was added in 1939. This became the sole power source in 1943 after the mill was struck by lightning and the sails were damaged. An electric pump made the diesel one redundant fourteen years later. The National Trust took the mill over in 1948 from the Buxton family, who continued to run the estate.

Over the intervening years Horsey mill has been restored at regular intervals. It underwent major restoration in 1959–62 and further work was needed following the great storm of October 1987, when the fantail was lost. The sails were removed for repair on 30 April 2014 and were expected to be absent for approximately eighteen months. At the time of writing the mill is still without its sails and fan but the eventual aim is to reinstate the sails and hopefully get them turning again.

Thurne Dyke Mill, Thurne

Location: Thurne Dyke, NR29 3BU. East bank of River Thurne at head of Thurne Staithe. Accessible by water or by footpath from road next to Staithe. Run by Norfolk Windmills Trust and open to the public on the second and fourth Sunday of the month from April to September, or by appointment.

This attractive and photogenic little tower windpump, which stands by the River Thurne at the opening to the Staithe, is another of Norfolk's most famous mills. When it was built in 1820 it had only two floors plus cap but was later hained (made taller) with the addition of a third storey. The point at which the work took place is marked by a black iron band on the white-painted tower, and the brickwork above rises in a straight line to the cap. This is thought to have happened around the middle of the nineteenth century and coincided with the change from common sails to the more modern patent type. It is probable that at about this time the mill gained a turbine pump in place of the original scoop wheel. Thurne Mill's cap and sails were blown off in 1919 but were

Thurne Mill, June 2010.

soon put back in place. It continued to work by wind power until just before the start of the Second World War, when further gale damage finally halted the sails. A steam turbine was installed in a shed close by in 1926, but the mill's future was uncertain before Mr R. D. Morse (see also Repps Wind Energy Museum, Chapter 4) rescued it for posterity in 1949. Restoration work by Mr Morse and millwright Albert England reversed its fortunes and the mill is now run by Norfolk Windmills Trust. Although always kept in good condition from then on, it had to wait until 2002 before being returned to working order. The work was carried out by millwright Vincent Pargeter. The windpump was later featured with sails turning in a television programme recording the travels of David Dimbleby. Visitors can enter at certain times during the summer months.

Horning – St Benet's Level Mill
Location: West bank of River Thurne, 0.25 mile north of Thurne Mouth. Accessible by water or on foot. From Ludham Bridge heading towards Ludham on A1062: turn right into Hall Road (NR29 5NU) and take second right turning after Ludham Hall, then at Cold Harbour take footpath by riverbank. Owned by Crown Estates and open to the public on the first Sunday in August, or by prior arrangement.

One of the earliest Broads tower windpumps, St Benet's Level Mill has had an eventful life and has undergone much alteration over the years. It was built around 1775 as a three-storey tower with a top floor being added at the close of the nineteenth century. The most radical change was a short lived conversion to an American-style windpump, involving the installation of an annular (circular) sail with a diameter of around 40 feet (12.2 metres). There is some debate as to exactly when this happened but according to one report it only lasted from 1896 to 1898. Mystery still surrounds the reasons for the conversion and the swift return to a traditional set of sails, though it is likely that the annular sail was badly damaged or blown down during a gale.

St Benet's Level Mill, June 2013.

After extending the brickwork at the top of the tower by an additional 10 feet (3.05 metres), Ludham millwright Daniel England fitted a boat-shaped cap with a highly unusual ten-bladed fantail. The added top section was tarred, in contrast to the rest of the brickwork. The new sails were the modern patent type, instead of the original old-fashioned common sails. A new turbine pump was probably installed at the same time. Like St Benet's Abbey Mill, it is officially in Horning but is several miles from the village of Horning. It is closer to Ludham and stands by the River Thurne almost opposite the much more famous Thurne Dyke Mill. It was restored by millwrights Lennard & Lawn between 1973–75 and again in 1989–90 by Richard Seago. It is not in working order but still retains its cap, skeleton sails and fantail. It can be reached by boat or on foot but is not generally open to the public.

Potter Heigham – High's Mill and Repps Level Mill

Locations: High's Mill: Potter Heigham, NR29 5NE. West bank of River Thurne. Repps Level Mill: 48 Riverside, Repps with Bastwick, NR29 5JY. East bank of River Thurne. Both accessible by river or by footpaths from Potter Heigham Bridge. Both privately owned and not open to the public. Potter Heigham Bridge is off A149 between Stalham and Caister-on-Sea or through Potter Heigham village heading east from Hoveton (A1062).

At Potter Heigham Bridge, a small riverside settlement just outside the village of Potter Heigham, footpaths lead to two former drainage mills. Standing within a mile of each other on opposite banks of the river, High's Mill and Repps Level Mill were typical examples of Norfolk Broads windpumps but both have been converted to residential accommodation. They both have caps but lack sails and fantails.

Left: High's Mill, Potter Heigham, March 2013. *Right*: Repps Level Mill, July 2012.

High's Mill was built by William Rust, who also erected Hunsett and Turf Fen mills on the River Ant. It is a fifteen-minute walk taking the pathway by the side of a public house on the west bank of the River Thurne and continuing past a long line of holiday bungalows. Though the date 1805 appears in metal numerals on the four-storey tower, it is thought to have been built around seventy years later. Tall trees shield the mill from close scrutiny during the summer months but are a less effective form of camouflage in winter. During its working life it was more clearly visible and probably retained its four sails until the 1920s or `30s. Formerly known as Grapes' Mill after the family who ran it for many years, it became known by its present name after the High family took charge in 1924. A steam engine supplemented wind power and continued operating till 1938. The mill was sold at auction in the 1990s and was advertised for sale again in May 2015.

On the east bank of the river and in the opposite direction to High's Mill, another path leading from Potter Heigham Bridge passes the black-painted Repps Level Mill. It can be seen at a distance from the ancient stone bridge and is roughly a five minute walk, again passing a row of holiday dwellings. Though out-of-bounds to the public, it can be viewed at close quarters from the track that runs by the river. For many years it lacked a cap but a new boat-shaped one was fitted in 2010. It is surprising how much more imposing the tower now looks with its replacement cap, but it would look even better if it had sails! In the summer of 2013 it was offered for sale and is currently available to rent as holiday accommodation. On the other side of the river, almost exactly opposite Repps Level Mill, stands an unusual white holiday home constructed from a fairground attraction which once stood on Great Yarmouth seafront in the early twentieth century.

RIVER ANT

Stalham – Hunsett Mill

Location: Chapel Field, Stalham, NR12 9EL. East bank of the River Ant, approx.
1 mile southeast of Wayford Bridge. Not accessible by road. Privately owned and not
open to the public.

This charming three-storey drainage mill may not be a household name but its
image has appeared numerous times over the years on picture postcards, calendars and
various other items of Broads-related merchandise. It is situated in private grounds at
the end of a long track, though the iconic 'chocolate box' view can only be appreciated
from the River Ant. Dating from the 1860s Hunsett Mill's tower has a stone bearing
the date 1698, which may relate to an earlier windpump known to have existed on
the site. The present mill was constructed by local millwright William Rust and is one
of just three in Norfolk known to have had two scoop wheels instead of just one (see
also Turf Fen Mill). The twin scoop wheels were positioned to the left and right hand
sides of the tower and could also be driven by an auxiliary engine. Four patent sails,
a six-bladed fantail and a white boat-shaped cap topped the red-brick tower. Once
associated with the Pratt family, it had been idle for a long time when restoration began
in 1970. Since then the mill has had several different owners. A sail was replaced due to
gale damage in 1989 and the fantail was blown down in 2007. A new one was installed
the following year. Externally Hunsett Mill still looks in excellent condition with its
cap, fantail and skeleton sails all in position, but most of the internal machinery was
taken out long ago. In 2009 a new 'shadow' extension was built onto the nineteenth
century mill house after earlier additions to the property were demolished, and won a
prestigious architectural award the following year.

Hunsett Mill, 1999. (J. Neville)

Irstead – Turf Fen Mill (opposite How Hill)

Location: Irstead, NR12. West bank of the River Ant opposite How Hill Nature Reserve. Not accessible by road. Visitors can park at How Hill (NR29 5PG; off A1062 approx. 1.75 miles west of Ludham) and follow riverside path on the east bank. Owned by Norfolk County Council and not open to the public.

Like Hunsett Mill, this restored windpump at Turf Fen was also the work of William Rust and featured twin scoop wheels. These could be operated together or independently. The tower only has two floors but stands 31 feet (9.45 metres) tall plus its boat-shaped cap, six-bladed fantail and four double-shuttered sails. It was built during the 1860s or `70s to drain water from the Horning marshes and probably continued to fulfil this role until the 1940s. After being taken into the care of Norfolk Windmills Trust in 1976 it was later restored from a derelict state by millwright John Lawn. Major repairs took place during the early and mid-1980s and were complete by 1986. The work included fitting a new cap, a new set of sails and a specially built replacement head wheel. Today it is still in fine condition but is not operational. Turf Fen Mill is not accessible to the public and can only be reached by boat, though it can be clearly seen from the How Hill Nature Reserve on the east bank of the River Ant.

Visitors to How Hill can also walk to Clayrack and Boardman's mills (see Chapter 4), two small windpumps roughly 200 yards (182.8 metres) apart. A converted corn mill (see How Hill Mill, Chapter 1) can be seen close to the main car park.

Above and below: Turf Fen Mill, Irstead, June 2013.

Ludham Bridge North Mill, Ludham

Location: East bank of River Ant just north of Ludham Bridge, NR29 5NX, on A1062 between Horning and Ludham. Stands beside footpath from Ludham Bridge. Empty tower not open to the public.

Standing just 20 feet (6.1 metres) tall and only 12 feet 4 inches (3.75 metres) in diameter at the base, the derelict shell of Ludham Bridge North Mill still exists amid bushes and undergrowth. A small community has built up around the bridge and is separate from the village of Ludham a mile-or-so away. The mill was built in the nineteenth century and by 1934 had parted company with its sails and fantail but still retained a cap complete with gallery. That was later removed and during the Second World War it was taken over by the Home Guard and turned into a pillbox. Evidence of their occupation can still be seen in the form of loopholes in the tower and the remains of a concrete mortar mounting block. This can be glimpsed on the right of the photograph in front of the doorway. In its working days it was a typical example of a small Broads windpump but its location makes it more accessible than most for the modern-day enthusiast. In summer a long line of hire boats extends from the bridge well past the lone tower. A few years ago it was possible to stand inside the remains of the tiny mill, but for safety reasons visitors are no longer allowed to enter and it has become rather overgrown. Now topped by an open wooden frame the tower has developed a distinct lean over the years, though it attracts far less attention than a more severely afflicted (and much more famous) tower in Pisa.

Another windpump known as Ludham Bridge South Mill or Beaumont's Mill once stood to the south of the bridge but was removed to create a mooring area for boats.

Ludham Bridge North Mill, March 2013.

Chapter 3

Post Corn Mills

Norfolk once had hundreds of post mills grinding corn and other grains. Today it has just three complete or near complete examples but none are currently operational. One (South Walsham) did not exist until the last decade of the twentieth century and has not yet been completed. Of the other two, only one (Garboldisham) has its original (but extensively rebuilt) roundhouse and buck (the main wooden body which revolves to face the wind direction), while the other (Thrigby) has a completely new buck perched atop the original roundhouse and is the only one of the three to have sails. The ruins of Topcroft Mill still exist plus a few other scattered fragments elsewhere. This chapter takes a closer look at all of the post mills named above. Space does not allow descriptions of Norfolk's many 'lost' post mills of which no trace remains, but they are represented here by two of the best known, Sprowston Mill and Swanton Morley Mill.

Garboldisham Post Mill, near Diss
Location: Hopton Road, Garboldisham, IP22 2RJ. South side of B1111, off A1066, approx. 1 mile south of Garboldisham village and 6 miles west of Diss. Privately owned and opens to the public by prior arrangement only.

As 1778 is the earliest date recorded inside Garboldisham Post Mill, it is likely that it was constructed in that year or a little earlier by James Turner. It may have originated as an open structure with the brick roundhouse possibly being a slightly later addition. Major improvements in around 1830 included the installation of new patent sails (replacing the old common type), an eight-bladed fantail and new iron gearing. Two pairs of millstones were positioned in the wooden body or buck above the roundhouse. After two sails were lost during a gale in August 1879, a new set of wider sails were fitted by Alfonso Vincent. The tail of the buck – the county's longest measuring nearly 22 feet (6.7 metres) in length – was extended at the same time. The mill was reduced to just one pair of sails and a single stock following a storm in 1906, when one sail was blown off and its opposite number was removed. It continued to operate in this fashion but, following the sudden death of miller William Bennett in 1914, seldom if ever worked again. Christopher Pattinson, a miller and farmer who had employed Mr Bennett, did not renew his lease on the mill when it expired three years later.

Garboldisham Post Mill was derelict when George Colman paid £250 for the sad remains in 1972. His son Adrian and others began a long restoration programme that

Garboldisham Post Mill, March 2011. (A. Colman)

would bring the mill back from the edge of oblivion. They replaced the decayed wooden trestle and rebuilt the brick roundhouse. Originally a two-storey design, it was rebuilt with just one floor. The buck was restored between 1975–78, which involved replacing the weatherboarding and repairing the floor and roof. An electric mill was installed in the roundhouse around 1975 and much additional work between 1979 and 1997 included replacing the external stairway and fantail. Some of the gearing and machinery was salvaged from donor mills. The weatherboarding was painted white with contrasting black roofs for the roundhouse and buck. To date the sails have not been replaced but otherwise the mill looks complete externally. In addition to the post mill, a smock mill and a four-storey tower mill also once existed on what was then common land. The former was constructed by James Turner about 1788 but may have disappeared by the end of the 1830s, while the latter was built in 1820. According to one account the tower mill was gutted by fire in August 1840, though it may have remained standing for another twenty years.

Thrigby Post Mill, near Caister-on-Sea

Location: Mill Road, Thrigby, NR29 3DY. 1 mile east of Thrigby village, approx. 3 miles west of Caister, off A1064. Privately owned and open to the public on Bank Holiday Mondays and some other days or by prior arrangement.

Only the original brick roundhouse dating from the 1790s still survived when new owner Nick Prior embarked on an ambitious quest to restore Thrigby Post Mill to its former glory in the early 1980s. After repairs were made to the roundhouse, which has an external diameter of 25 feet 6 inches (7.77 metres), millwright John Lawn installed a new centre post and carried out other work in early 1983. A brand-new buck was made and put in place on top of the thatched two-storey roundhouse by 1985. New canvas covered common sails were fitted in 2004. The interior was equipped with two pairs of millstones with the eventual aim of Thrigby becoming a working mill again. Mr Prior sold the mill to Peter Gillett in 2007 and since then it has been open to the public on National Mills Weekends and other selected dates.

Thrigby Post Mill, April 2013.

Thrigby Post Mill is known to have existed in 1797 and may have been built between five and seven years earlier. It was erected for Robert Woolmer who resided at Thrigby Hall and needed a mill to grind wheat for his estate. A succession of tenant millers passed through its doors over the course of almost a century, the last being Alfred C. Hood. The mill stopped work in 1889 and the buck was taken down three years later due to death watch beetle infestation. The roundhouse was left to decay before restoration began ninety years later.

South Walsham Post Mill

Location: Mill Lane, South Walsham, NR13 6DF. Off B1140 between South Walsham village and Acle. Privately owned and not open to the public.

The first completely new post mill to be erected in Norfolk in well over a century, this impressive replica was built by millwright Richard Seago and has become a local landmark about a mile east of the picturesque village of South Walsham. Work commenced in 1994 on a large brick roundhouse, and in November 2000, a wooden body or buck was lifted into place by crane. The main wooden post is 20 feet (6.1 metres) long and 2.5 feet (0.76 metres) square at the base. Some of the ironwork was salvaged from lost mills. A tail ladder was recently added which can be clearly seen in the photograph. The owner's intention is to fit sails and a fantail and eventually run the mill on a semi-commercial basis. South Walsham once had an earlier post mill close to the site of the newcomer, built circa 1759 and demolished around 1870. A tower mill in nearby Flowerdew Lane was also in existence from the 1790s to 1878.

Topcroft Post Mill, near Bungay

Location: South of the B1135, Topcroft, NR35, northwest of Bungay. The substructure still exists but is privately owned and not accessible to the public.

Topcroft Mill was built before 1838 and John Rounce was the first recorded miller. At least eleven other millers came and went over the years before it stopped work around 1922. The sails and fantail were taken down in 1924 after the latter was damaged. Parts of the mill were later reused at other mills. During its working life Topcroft Mill had four double-shuttered sails, a six-bladed fan on the ladder, and drove two pairs of French Burr stones. The roundhouse was two storeys high and the central wooden post was 25 inches (0.64 metres) square at the base and 14 inches (0.36 metres) square at the top.

The entire substructure of Topcroft Mill still exists including the central post and trestle. Part of the brick roundhouse also survives but is overgrown. Nothing remains of the wooden buck.

South Walsham Post Mill, April 2013.

Topcroft Post Mill, 1910. (J. Neville)

Norwich – Sprowston Post Mill

Location: Mill Road, off Sprowston Road, Norwich, NR3. Mill destroyed by fire 1933.

One of Norfolk's most famous and best documented post mills, Sprowston Mill on the edge of Norwich was probably constructed in 1780 as an open skeleton or trestle mill, though much earlier dates of 1702 and 1730 have also been mooted. It was later rebuilt as an enclosed post mill with roundhouse and was steadily improved over the years. Thanks partly to its location on the edge of Mousehold Heath (a large area of woodland and heath on the outskirts of the city), the mill became a popular subject for artists and photographers alike. Thomas Carlton, who had been the miller since at least 1817, sold it in 1823 to Robert Robertson. He continued to work Sprowston Mill until 8 March 1842, when he was tragically killed at the age of forty-eight after getting caught in the machinery. The ill-fated Mr Robertson had been responsible for a number of technical improvements at the mill, including the installation of new patent sails and a fantail.

For many years Sprowston Mill was owned by members of the Harrison family, who grew their own wheat on farmland leased from the Gurney family of Sprowston Hall (now a hotel). Harrison's Corn Stores on Sprowston Road sold flour and other produce. William Harrison Jnr was the last owner before the mill itself met with a sudden demise on 24 March 1933. It is thought that someone was burning rubbish on Mousehold Heath and that this started a gorse fire that quickly spread and set light to the canvas on the mill's sails. On this occasion nobody was killed but, despite the best efforts of the fire brigade, little remained of the historic building and it was not rebuilt. Somewhat ironically, it was reported that the mill was due to have been taken over by Norfolk Archaeological Society on the day after it burned down, with the intention of saving it for posterity. The photograph reproduced here was probably taken less than a decade before it caught fire. Now just a representation on the town sign and a track off Sprowston Road marked Mill Road serve as reminders. The land where the mill stood is now occupied by a private dwelling and the surrounding area has been urbanised.

Sprowston Post Mill, *c.*1925. (M. Roots)

Given the horrific death that occurred at the mill it may not be entirely surprising that the pathway to it is allegedly haunted, though the ghost is said to be that of the last miller. Sadly, the disappearance of Sprowston Post Mill robbed Norwich of yet another of its once plentiful windmills. According to a map dated 1817 two other post mills also existed on Sprowston Road.

Swanton Morley Post Mill, near East Dereham

Location: West side of the B1147, Swanton Morley, NR20, northeast of East Dereham. All remains have now been demolished.

Swanton Morley Mill boasted the tallest roundhouse of any known Norfolk post mill. It had three floors, stood 20 feet (6.1 metres) tall to the eaves and had a base diameter of 26.5 feet (8.1 metres). It is known to have existed in 1795 and had many different millers during a working life of more than 100 years. Patent sails and an eight-bladed fan on the rear ladder were probably fitted during the 1850s. The mill drove two pairs of stones by wind and a further pair by auxiliary power. A Crossley paraffin engine was in use by 1904 in lieu of a portable steam engine that had been *in situ* since at least 1891. The sails were blown off during a fierce gale in 1895 and the buck (upper section) was dismantled in 1906 by a carpenter named Joseph Canham, who later made a large model of the mill which still exists. Nothing now remains of Swanton Morley Post Mill. The roundhouse was still in use as a store room until at least the late 1980s but was finally torn down as recently as 2008 to make way for housing.

The second tallest roundhouse in Norfolk belonged to Holt Heath Post Mill on Hunworth Road, Holt (NR25 6SR), and part of the wall still survives though it is overgrown. It had three storeys and was 18 feet (5.49 metres) tall to the eaves and 24 feet (7.31 metres) in diameter at the base.

Swanton Morley Post Mill, *c.* 1906. (J. Neville)

Chapter 4

Rare and Unusual Mills

COMBINED WIND- AND WATER MILLS

In addition to its numerous windmills and windpumps, Norfolk was also home to a number of water-powered mills. The latter are obviously beyond the scope of this book but two curious hybrids still exist in the county. They are far from identical but fulfilled the role of combined wind- and watermill in different ways. One consists of a full-sized watermill with a windmill attached while the other is a tower windmill with a separate waterwheel and pumphouse. Though neither is in working order they are both of historical importance and well worth viewing.

Burnham Overy – Union Mills Combined Wind- and Watermill
Location: Off A149 at Burnham Overy, on River Bure, PE31. Privately owned and not open to the public.

Union Mills, Burnham Overy, March 2012.

In addition to Savory's Mill (see Chapter 1) Burnham Overy is probably unique in also boasting an impressive watermill (owned by the National Trust and converted to residential use) and an unusual combined wind- and watermill. Known as Union Mills, the latter started life as a conventional watermill before a tall windmill was built adjoining it in 1814. Erected by Thomas Beeston, the miller on the site from 1802 to 1825, the windmill stands six storeys tall plus cap, but appears to originally have had five floors. The watermill, claimed to date from 1737, has also been heightened during its long lifetime. Thomas Beeston's initials, the date 1814 and the word 'peace' are still to be seen above the windmill's doorway today. It once had four double-shuttered patent sails with copper vanes. Different sources claim that the windmill had either two or three pairs of French Burr millstones and the watermill between three and four pairs. Both mills were capable of running the other's stones by either wind or water power and were fully interconnected internally.

After failing to sell at auction in Norwich in August 1825, a miller named James Read operated both mills between 1830 and his death in 1864, when they were leased to his son-in-law, William Love Porritt. He continued working until the turn of the century, though the mills were put up for auction again in 1870. They were acquired in 1904 by Philip Roy, who later set up a limited company with his sons, and were worked by members of the Roy family for many years. The windmill stopped working by wind power in 1893 and its machinery was sold, but the watermill continued to operate using roller milling equipment. A steam engine was in use during the 1890s and was superseded by gas and oil engines. The tower was badly damaged by fire in May 1935 but was repaired. The mills were flooded in 1953 but continued to be used until the end of the 1960s. Both sections of Union Mills still exist and the watermill still has some of its machinery. Standing on the River Bure about a mile from Savory's Mill in an attractive location, it is not as well-known as its neighbour but still serves as a local landmark. After being bought privately in 1999, both buildings were extensively renovated and the windmill tower was converted into living accommodation. It has a new ogee-shaped cap and reefing stage but no sails.

Union Mills, Burnham Overy, March 2012.

Little Cressingham Combined Wind- and Watermill, near Watton

Location: Little Cressingham, IP25 6NT. Approx. 3 miles west of Watton on B1108. Owned by Norfolk County Council and run by Norfolk Windmills Trust. Exterior accessible at all times but not currently open to the public.

Unlike Union Mills, Little Cressingham at first glance looks like a conventional tower windmill. On closer inspection it soon becomes apparent that there is a waterwheel connected to it and a pumphouse a short distance away. Built about 1820–21 on the Clermont estate, the six storey tower stood 50 feet (15.24 metres) tall (plus cap) and housed four pairs of stones. Two pairs in the top half were powered by the wind and sails while the other two pairs in the lower half were driven by the 12 feet (3.66 metres) by 6 feet (1.83 metres) waterwheel. An additional smaller waterwheel in the pumphouse enabled a Bramah pump to supply water to Clermont Lodge about a mile distant.

Sadly, a tragedy occurred in the nearby mill house on the night of December 25 1890, when thirty-six-year-old miller Samuel Goddard and his wife Elizabeth (thirty-seven) were suffocated due to fumes from hot coals that they had taken to their bedroom. The mill continued to be worked by wind power until 1916, when it was tail-winded and badly damaged. One pair of sails had been blown off five years earlier. The cap was removed during the 1940s and was replaced by a flat roof. Nevertheless, commercial milling continued onsite up until 1952 by means of the waterwheel, an oil engine and a roller mill. The lease of Little Cressingham Mill was transferred to Norfolk Windmills Trust in 1981 and restoration work started later that decade and continued into the 1990s. Between 1989 and 1991 the pumphouse, Bramah pump and the main waterwheel were repaired. New windows were fitted in the tower in 1993. Today the mill is in a good state of repair but still lacks a cap and sails. Norfolk Windmills Trust

Little Cressingham Combined Mill, September 2012.

Left: Little Cressingham Mill waterwheel, September 2012. *Right*: Little Cressingham Mill Pump House, September 2012.

has not ruled out the possibility of these being installed at some point in the future. Located close to the road on a green, the mill can be viewed at close quarters at any reasonable time but is not currently open to the public.

SMOCK MILLS

Relatively rare in Norfolk compared with tower and post mills, the smock mill consisted of a wooden body or smock placed on top of a base usually made from brick or stone and was cheaper to produce than an all brick tower. The base was often quite low but could be as much as three or four storeys tall. Some very tall smock mills still exist in the Netherlands and the USA. England's tallest surviving smock mill is Union Mill at Cranbrook in Kent, built in 1814 and 72 feet (21.94 metres) high including the cap.

It is often stated that smock mills were so called due to a perceived visual similarity to the garment of that name once worn by countrymen and boys. However, the late Harry Apling, a respected mills expert and author, thought it more likely that they were named after an earlier article of women's underwear also known as a smock, which was popular for hundreds of years until the close of the eighteenth century. As with tower mills, the smock variety was divided into traditional corn mill and windpump categories. There are no complete unmodified smock mills left in Norfolk but a few examples still exist in various forms. The most significant of these are detailed below.

Horning Ferry Smock Mill, Horning
Location: Ferry Road, Horning, NR12 8PS. Off A1062. Visible from the road and the River Bure but privately owned and converted to residential use. Not open to the public.

Horning Ferry Mill, June 2013.

Horning Ferry Mill as it appears today bears little visual resemblance to the smock windpump built by England's of Ludham. In its working days the three storey, octagonal, weatherboarded mill was tarred, while the cap, fantail and sails were painted white. Its year of build is unknown with estimates ranging from the 1870s to around 1900. In 1935 a highly unusual residential conversion by H. P. E. Neave saw the mill transformed almost into a fantasy version of its former self, complete with sloping sides and a large single-storey extension encircling the base. To accommodate this extension the sails were replaced with smaller 'dummies' but the cap, gallery and seven-bladed fantail survived. Much of the original mill remains within the new facade, including some of the machinery. Seen to its best advantage from the water, the end result is a unique building which still intrigues holidaymakers and its image has been published numerous times. During the 1960s the sail stocks and cap were painted red but apart from the stocks which are currently painted black the whole mill is now a vision in white!

Briningham Smock Mill, near Melton Constable

Location: Mill Lane, Briningham, Melton Constable, NR24 2QA. Off the B1354/ B1110. Forms part of the Belle Vue Tower. Privately owned and not open to the public.

Like Horning Ferry Mill, Briningham Smock Mill has also been incorporated into a private residence but there the similarity ends. Built in 1721 and thought to be the oldest surviving remains of a Norfolk windmill, the octagonal brickwork is three storeys and about 29 feet (8.84 metres) tall. It was retained as the base for a viewing tower when the original two-storey wooden upper section of the mill was taken down in the latter half of the eighteenth century. An additional three storeys of cylindrical brickwork was added and the Belle Vue Tower, as it is known, has a total height of approximately 50 feet (15.24 metres). It also has the distinction of being the first mill in the county to be converted to living accommodation, long before it became fashionable.

Belle Vue Tower, Briningham, July 2015.

Briningham Smock Mill was built at the behest of Sir Jacob Astley of Melton Constable Hall and probably ceased work during the 1770s before its dramatic transformation. Two centuries later, Belle Vue Tower was rescued from dereliction and restored during the late 1970s and early `80s. It is obscured by tall trees at the bottom of a long private driveway, but can be seen at a distance from various vantage points in the locality.

Apart from the above there is not much surviving evidence of smock mills in the county, though the preserved remains of Tunstall Smock Windpump (NR13) still exist. It was truncated to form part of an engine house and stands beside the River Bure some distance from any public road. The ten-sided base of Fakenham Smock Mill was incorporated into the bottom section of a private dwelling named Tower House in Holt Road, Fakenham, NR21. It was originally built in the early nineteenth century before being reconstructed as a tower mill around 1849 and finally becoming a residence during the twentieth century.

SKELETON/OPEN TRESTLE WINDPUMPS

Only three open trestle or skeleton windpumps still remain in Norfolk. As the name suggests they were simple wooden structures open to the elements and were utilised to drain areas where a larger tower or smock mill was considered unnecessary or too expensive. Due to their timber construction most have rotted into oblivion and the surviving trio have been extensively renovated or rebuilt over the years.

St Olaves Mill (aka Priory Mill), near Fritton

Location: St Olaves, NR31. East bank of the River Waveney just north of St Olaves bridge on the A143, approx. 1 mile west of Fritton. Accessible from the river or by footpath from the bridge and can be viewed externally at all times. Leased to Norfolk County Council and not generally open to the public.

The village of St Olaves is probably best known for the ruins of its priory but this little windpump dating from 1910 is also worthy of attention. It was built by Ludham millwright Dan England as an open trestle or skeleton mill and in 1928 was given a weatherboarded body. This modification changed the mill's external appearance considerably but it is still recognised as one of only three original skeleton drainage mills still in existence on the Norfolk Broads. St Olaves Mill (also called Priory Mill) replaced a smock mill that stood on the site until 1898. Until 1974 it was a Suffolk mill, the move into Norfolk being the result of border realignment rather than physical relocation. A small scoop wheel is still enclosed within the structure and it continued to work until 1957, when wind power gave way to electricity. It was renovated and brought back to working order over several years in the 1970s and early `80s. During its lifetime St Olaves Mill has fallen victim to strong winds a number of times, one of the most spectacular being in January 2007, when the cap and sails were toppled during a severe gale. Since then it has been repaired and once again has its miniature cap, fantail and four patent sails all back in position.

St Olaves Mill, September 2006.

Ludham, How Hill – Boardman's Mill

Location: How Hill Nature Reserve, NR29 5PG. On east bank of the River Ant, approx. 200 yards (182.8 metres) south of Clayrack Mill. Visitors can park at How Hill (off A1062, approx. 1.75 miles west of Ludham) and follow riverside path. Owned by Norfolk County Council and accessible to the public.

Named after Edward Boardman, who owned the How Hill Estate for many years (see also How Hill Mill, Chapter 1), this is another of the county's surviving trio of skeleton mills and it has remained unmodified. Like St Olaves Mill it was the work of Dan England and is thought to date from around 1897. Boardman's mill worked as a windpump – firstly with a scoop wheel and from 1926 utilising a more modern Appold turbine pump – for just over forty years before succumbing to terminal gale damage in 1938. After this it stood derelict for more than four decades before being restored in 1981 by the Norfolk Windmills Trust in association with the Broads Authority. Now Grade II listed it is nearly 30 feet (9.14 metres) tall including cap. It still has its replacement sails and fantail, and can be viewed at all times from the footpath that runs beside it.

The only other skeleton open trestle drainage mill left in Norfolk is Hobbs' Mill, which dates from the latter part of the nineteenth century. It is located on Ferry View Estate, Horning (NR12 8PT) but is not accessible to the public. The mill, which still retains its machinery, has a replacement cap and fantail but no sails. It stopped work during the 1930s and was restored five decades later by millwright John Lawn and others.

Boardman's Mill, How Hill, June 2013.

HOLLOW POST WINDPUMPS

Essentially a basic scaled-down post mill complete with a tiny timber body or buck, sails and fantail, the hollow post windpump was designed to pump water instead of grinding corn. Unlike a conventional post mill the central post contained a shaft or rod which (directly or by means of gears) operated a plunger pump or scoop wheel. Like skeleton mills, the hollow post type was a relatively lightweight and inexpensive alternative to a tower drainage mill, but was also at the mercy of the elements and prone to decay. Only three are known to still exist in Norfolk and these are examined below.

Ludham, How Hill – Clayrack Mill

Location: How Hill Nature Reserve, Ludham, NR29 5PG. On the east bank of the River Ant, approx. 200 yards (182.8 metres) north of Boardman's Mill. Other location information as Boardman's Mill.

The exact year of build of this highly unusual hollow post windpump is unknown but it probably dates from the second half of the nineteenth century. It originally stood on Ranworth marshes around 3 miles from its present location and worked until the early twentieth century. It was 'rediscovered' nearly eighty years later, by which time only fragments remained and it was necessary to reconstruct the buck, sails and fantail. The work was carried out by millwright Richard Seago on behalf of Norfolk Windmills Trust in the late 1980s. The decision to move the mill to How Hill for restoration was at least partly to avoid disruption to nesting birds at its original home. It fits in very well with Boardman's Mill a short distance along the riverbank and helps enhance the idyllic Broads scene. After being restored the mill was once more capable of pumping water using a scoop wheel. However, it has for some time been missing its sails but still has sail stocks.

Left: Clayrack Mill, How Hill, June 2013. *Right*: Palmer's Mill, Upton Dyke, July 2013.

Upton Dyke, near Acle – Palmer's Mill

Location: On field beside Upton Boat Dyke, off River Bure, NR13. Public car park off Boat Dyke Road, approx. 1 mile northeast of Upton village (follow signs to Boat Dyke) near Acle. Can be viewed at all times.

This charming little hollow post windpump is situated at Upton Dyke but was actually built a few miles away in Acle. Like Clayrack Mill it was extensively rebuilt after becoming derelict by Richard Seago. When I first saw Palmer's Mill, several years ago, I was visiting friends who were caravanning beside it but we were oblivious to its historical significance. It stopped working on its original site in the 1920s, only a couple of decades after starting. Following relocation to Upton Dyke the mill was renovated in the late 1970s and the start of the 1980s and is again in working order. The sails measure 21 feet 9 inches (6.63 metres) in diameter and drive a plunger pump in lieu of a scoop wheel. For those who enjoy a hike, Upton Black Mill on the River Bure is within walking distance and the tower windpumps at Oby and Clippesby can be viewed across the water (see Chapter 2).

Starston Mill, near Harleston

Location: Mill Field, Pulham Road, Starston, IP20 9NR. West of Starston village, approx. 330 yards (301.75 metres) southwest of St Margaret's church. Privately owned but accessible to the public at all times.

This recently restored hollow post windpump in south Norfolk close to the border with Suffolk was constructed around the middle of the nineteenth century by Whitmore & Binyon of Suffolk. Starston Mill differs from the two other hollow post windpumps already mentioned in that its base consists of a tiny brick roundhouse similar to but much smaller than those of corn post mills. The mill had no scoop wheel but drove a pair of

Starston Mill, July 2015.

Starston Village sign, July 2015.

plunger pumps. Water was pumped from Starston Beck to Starston Place for use at Home Farm. Its condition deteriorated over the years and it was placed on the Heritage at Risk Register before being restored in 2010. The attractive village of Starston is dominated by its church but the windpump is tucked away in a corner of a field off Pulham Road. It would be easy to miss if you were unaware of its existence. The windpump and church are depicted on the village sign dating from 1980. A corn post mill also existed in the village between about 1836 and 1879 and may later have been relocated to Suffolk.

WIND ENGINE COLLECTION

Repps – Wind Energy Museum (The Morse Collection)
Location: Staithe Road, Repps with Bastwick, NR29 5JU. From the A149: take Church Road past St Peter's church then turn right into Staithe Road. Follow signs to Wind Energy Museum. The museum is on private property and opens to the public on selected dates only.

Wind Engine is a term used to describe different kinds of devices which use wind power for various purposes. Usually much smaller and cheaper to produce and maintain than traditional brick windmills or windpumps, they were often used for drainage in areas where a larger mill was not needed and could also be employed to drive saws or other machinery. Many were wooden but later examples were often made of metal. They were particularly prevalent in the USA during the late nineteenth and early twentieth centuries, where they were used primarily to pump water. A very tall one at the XIT Ranch in Littlefield, Texas, was said to 132 feet (40.23 metres) high before being blown down on Thanksgiving Day in 1926.

A unique collection of wind engines can be found deep in the Norfolk countryside at Repps with Bastwick, close to the River Thurne. The late Mr R. D. Morse, who also

Repps Wind Energy Museum, August 2015: Windpump scoop wheel (*foreground*) and other exhibits.

helped save Thurne Mill (see Chapter 2) for posterity, built up his collection over the course of nearly half a century. Some were rescued from dereliction and restored to working order. Now named the Wind Energy Museum but formerly known as Morse's Wind Engine Park, the attraction occupies a secluded 2.75-acre site. The collection includes examples from North America, Australia and Great Britain. Some have annular (circular) sails (see also St Benet's Level Mill, Chapter 2) while others have miniature conventional sails. A windpump scoop wheel originally sited at Whitlingham Lane, Norwich, now entertains visitors using electricity instead of wind power. One of the oldest exhibits, a small black windpump with contrasting tiny white sails, dates from around the 1850s. A stationary steam engine provides additional interest. Models of various mills including Thurne can be seen in an indoor area. The museum is accessible to the public (charges apply) on a number of open days each year.

Repps Wind Energy Museum, August 2015: Monitor Windpump *c.* 1930s from Nebraska (*left*); Assorted Wind Engines (*right*).

Chapter 5

The Lost Giants: Norfolk's Largest Demolished and Truncated Mills

Norfolk is known to have had two of the tallest traditional windmills ever constructed anywhere in the world and for a while they worked simultaneously. The mammoth eleven-storey mills at Great Yarmouth and Bixley – one of which disappeared from the coastal skyline well over a century ago while the other still exists in truncated form – are examined in detail below. Reaching a claimed height of 137 feet (41.76 metres), Bixley Mill would have dwarfed even the world's tallest remaining conventional windmill, The North, which is located at Schiedam in Holland and stands 109 feet 3 inches (33.3 metres) high to the top of the cap. Britain's tallest surviving windmill at Moulton in Lincolnshire has been fully restored to working order and now just qualifies for the 100 feet (30.5 metres) high club thanks to a 3 feet (0.9 metres) high finial on top of its new cap.

Several other Norfolk mills were eight or nine storeys high and were taller than any still in existence in the county. In the twenty-first century it seems amazing that these massive structures simply disappeared without trace. Unlike many of today's survivors the lost giants did not have the protection of listed building status. They are arranged in approximate height order starting with the tallest.

Bixley Mill, near Norwich
Location: Bixley Manor, Kirby Bedon Road, Bixley, NR14 8SJ. Just off the A146, approx. 2 miles southeast of Norwich. Converted remains in private grounds. Not accessible to the public and not visible from the highway. Truncated *c.* 1865.

Unlike most other mills in this chapter the lower section of Bixley Mill's tower still exists, though it is a shadow of its former self. Standing close to busy roads in the parish of Bixley to the south of Norwich, it is invisible to the public behind tall, impenetrable woodland in the grounds of Bixley Manor. There is nothing to suggest that here stand the remains of one of the tallest traditional windmills ever built.

The base of Bixley Mill is reputed to measure 43 feet (13.1m) in diameter with 42 inch (1.1 metres) thick walls, making it one of the widest mill towers ever recorded. No photographs exist prior to its unfortunate partial demolition in 1865, though detailed drawings from contemporary letterheads have survived. An early one is reproduced here showing the windmill and mill house before a steam mill was added. This gives some idea of the huge scale of the mill during its working days and similarities with

Southtown (Cobholm) High Mill at Great Yarmouth are evident. It is estimated that the original height was an amazing 137 feet (41.76 metres) to the top of the cap and 127 feet (38.7 metres) to the curb (top of the brickwork). The diameter of the sails is claimed to have been around 112 feet (34.14 metres). All this would suggest that Bixley Mill was truly a giant among giants. From the enormous sails to the height, diameter and thickness of the tower, it was constructed on a massive scale.

Built for Charles Clare in 1838 the mill had eleven floors plus its cap and housed five pairs of millstones. The tower was encircled by a reefing stage, which allowed the miller to adjust the sails with the aid of chains hanging down from the cap. Following a lightning strike on 23 May 1854, Thomas Smithdale of Norwich was called upon to make a new sail '13 yards long and 8 feet wide'. A steam mill was added at the site around 1857 with the installation of a steam engine built by Smithdale. Even with the combined might of wind and steam power, the Bixley Mills Company failed to make a profit and several unsuccessful attempts were made to let the properties as a going concern between 1863 and 1865. Finally, after a relatively short working life of just over a quarter of a century, the drastic decision was made to partially dismantle the windmill and the tower plus its machinery was sold at auction in 1865. Reduced to a still substantial seven storeys, it was purchased by George Levick Colman for £840 in July 1872. Later it passed into the ownership of Jeremiah James Colman and is still in the same family today. It was converted to a water tower and fulfilled this slightly ignominious role for many years. The adjacent millhouse survived intact and was leased separately when the mill was sold. It is still part of the estate today beside the mill which once cast a much longer shadow over it. Now converted to living accommodation the tower is still taller than many despite the loss of its four upper storeys and cap. It is strictly out of bounds to the general public.

Bixley Mill letterhead sketch, *c.* 1850
(M. Roots).

In addition to being the tallest mill ever built in Norfolk, Bixley was possibly also Britain's loftiest. A mill at Pickering, North Yorkshire, was advertised for sale in 1841 as having thirteen floors, though lack of additional information makes validation of this claim impossible. The eleven-storey Roe's Distillery Mill in Dublin, Republic of Ireland, may have been even taller, with claims of between 120 feet (36.58 metres) and 150 feet (45.7 metres). It lost its sails long ago but still exists and is now known as St Patrick's Tower. The Netherlands also had several giant mills taller than any that still survive there today. In terms of sail diameter Bixley was probably the largest in Europe, though it appears to have been narrowly beaten by the Murphy Windmill in California, USA, which had sails spanning a huge 114 feet (34.75 metres).

Great Yarmouth – Southtown (Aka Cobholm) High Mill

Location: Gatacre Road, Off High Mill Road, Cobholm, Southtown, Great Yarmouth, NR31 0BQ. Demolished in 1905.

Built in 1812 this giant was variously known as the Southtown, Cobholm or Great Yarmouth High Mill. It was situated close to the boundary between the Southtown and Cobholm districts of Great Yarmouth and was also referred to locally as Press's Mill after the milling firm Press Brothers, who owned and ran the building for many years. Like Bixley Mill it stood eleven storeys high plus cap. Various heights have been claimed including 102 feet (31.1 metres) to the curb (top of the brickwork) and either 112 feet (34.14 metres) or 122 feet (37.19 metres) including the cap. An iron cage 10 feet (3.05 metres) high attached to the top of the cap may explain the height discrepancy evident in various accounts. Claims of up to 135 feet (41.15 metres) to the top of a large lantern fitted in this cage have also been made. The whole ensemble was topped by a wooden weathervane that was later preserved in a local museum before being destroyed by fire during the Second World War. Being so tall this mighty mill was used as a navigation aid by vessels off the coast of Great Yarmouth. In its early days the High Mill appears to have almost had a dual function as a lighthouse but the lantern was soon extinguished as it was considered a fire risk.

The external diameter of the tower at ground level was reported to measure between 40 feet (12.19 metres) and 46 feet (14 metres) and the walls were 3 feet (0.91 metres) thick at the base. The tower originally housed four pairs of millstones (later increased to seven) and was wide enough for horse-drawn carts to enter for loading and unloading. The diameter of the four sails was 84 feet (21.51 metres), well behind the 112 feet (34.14 metres) span attributed to Bixley Mill and around the same size as those fitted to the much shorter Berney Arms Mill (see Chapter 2). It had a reefing stage encircling the tower at the sixth floor level and a gallery around the outside of the cap. A steam engine was later added to produce supplementary power. Estimates of the cost of the mill's construction vary widely depending on source, with figures of £2,000, £7,000 and £10,000 all having been quoted.

At its peak the mill worked around the clock with four men on each shift and was said to be Britain's most prolific windmill. During the Crimean War, flour produced

Southtown High Mill, Great Yarmouth, *c.* 1900.
(J. Neville)

here found its way to the British Army by means of clipper schooners. The High Mill was one of the most famous, best documented and most photographed windmills of its day and featured in period postcards of Great Yarmouth. The historic photograph reproduced here gives some idea of its sheer scale and presence. It was taken around the turn of the century after the mill had stopped working but a few years before it vanished from the town's skyline.

Southtown High Mill and Bixley Mill operated simultaneously for around twenty-seven years before the latter was literally given the chop! The High Mill had a much longer working life of eighty-two years and would probably have continued to grind corn into the twentieth century had it not been badly damaged by a lightning strike in 1894. It never worked again and was eventually sold at auction ten years later for a mere £100 before being completely demolished during 1905. The sails were reused at another of Press Brothers' properties, Roughton Mill (a few miles south of Cromer), which was only about half as tall as the donor mill. It still exists but long ago parted company with its oversized High Mill leftovers!

The High Mill was the tallest of several huge Great Yarmouth corn mills. More than a century later it seems a tragedy that the town has lost these iconic structures, which today would be major tourist attractions. However, the golden age of wind power was drawing to a close and the cost of repair and maintenance would no doubt have been as colossal as the mills themselves. The High Mill's tower contained an estimated 300,000 bricks, which were reused to construct a row of houses known as High Mill Terrace. These still stand on what is now Gatacre Road and serve as a direct link with the amazing building which once stood there. Nos 35 and 36 are on the actual site of the mill and were given red chimney pots in contrast to the plain stone colour of the other houses. A public house situated a short distance away in High Mill Road is appropriately named The Windmill.

Until relatively recently considered by many to have been the loftiest mill ever built in Norfolk, it has also been variously described as the tallest in Britain, Europe or the world. It may well have been one or all of these things at some point during its history. Cobholm High Mill was undoubtedly one of the tallest ever built but suffered the indignity of being eclipsed in its home county!

Great Yarmouth – Green Cap Tower Mill

Location: West of Lady Haven Dyke, Cobholm, Southtown, Great Yarmouth, NR31. Demolished *c.* 1905.

Green Cap Mill was another Great Yarmouth giant and was said to have been nine storeys high. As the name implies its cap was painted a slightly more unusual colour compared with the familiar white or black caps of most other mills. It had four patent sails and was located not a great distance from Southtown (Cobholm) High Mill. Like its even loftier neighbour, Green Cap Mill was later owned by Press Brothers but was built about 1815 for James Jenner. He advertised it for sale by auction at the Star Tavern, Great Yarmouth, in September 1831, along with a house and land. A granary adjoined the tower and there was even a pigeon loft – a rather novel feature for a working windmill. Sadly, many birds were probably burned to death when the mill was engulfed by fire on 29 January 1898. It was destroyed in the space of two hours with the under-equipped local brigade powerless to stop the fire's spread. The poignant photograph reproduced here shows the charred remains of Green Cap Mill with the doomed High Mill still standing in the background.

Green Cap Mill, Great
Yarmouth, 1898 (after the fire).
(J. Neville)

Great Yarmouth – South Middle Denes Tower Mill No. 6d

Location: Between Regent Road and Albion Road, Gt. Yarmouth, NR30. Demolished 1881.

This enormous mill once stood close to Great Yarmouth sea front, an area now at the heart of the town's tourism industry. It had nine floors, four double-shuttered patent sails and a boat-shaped cap complete with fantail. The tower was encircled by a reefing stage on the fourth floor and contained three pairs of stones. As the unusually precise name may suggest, Great Yarmouth South Middle Denes Tower Mill No. 6D was one of several operating in the locality during the nineteenth century. It was erected around 1824 and had an adjacent mill house, granary and other outbuildings. After failing to attract a buyer when offered for sale the mill was dismantled in 1881. Almost certainly the second tallest mill ever built in Great Yarmouth – though Green Cap Mill (see above) had the same number of floors – it was also one of the tallest ever built in Norfolk. Whether its height matched or surpassed the Upper Hellesdon and Bracondale mills in Norwich (see below) is unclear. Images of this mill are extremely rare but it deserves recognition as one of Norfolk's true giants. The only large sails on the sea front today are the dummy ones on the facade of the former Windmill Theatre – now housing a miniature indoor golf course – which serve as a reminder that milling was once a major Great Yarmouth industry.

Norwich – Upper Hellesdon Press Lane Mill (aka Witard's Mill)

Location: Press Lane, Upper Hellesdon, Norwich, NR3 2JY. Demolished 1920.

This Norwich colossus took over from Southtown (Cobholm) High Mill as the tallest surviving windmill in Norfolk and held the crown for about fifteen years before it too bit the proverbial dust. It was built in 1875 on the site of two earlier post mills and stood nine storeys high plus cap. As is often the case there is some disagreement as to how tall it actually was. An impressive claim of 95 feet (28.96 metres) to the curb of the tower plus the cap is contradicted by an alternative figure of 88 feet (26.82 metres)

Upper Hellesdon Mill, *c.* 1906 (during sail repainting).
(R. Bunn)

including cap. If the higher claim is exaggerated the lower one would still make it taller than all but one of Britain's remaining mills. The photograph shows men standing on the sails while painting them. Also known as Witard's Mill after Ephraim Witard, the miller who worked it from new till the sails finally stopped turning, it was damaged during a thunderstorm in 1906 but was repaired. Worse was to follow on 4 May 1913, when fire gutted the lower part of the tower and the roller mill adjoining it. An electrical fault was blamed and the structure stood idle for seven years before being pulled down. The bricks were reputed to have been reused in the construction of council houses, again drawing parallels with Southtown High Mill. According to one account, two of the sails found their way to the much smaller Stow Mill, Paston (see Chapter 1), where they had to be cut down to the required size. After Upper Hellesdon Tower Mill vanished from the skyline in 1920 its title of the county's tallest survivor was inherited by Sutton Mill near Stalham (see chapter one).

Norwich – Lakenham Bracondale Mill

Location: Bracondale, Lakenham, Norwich, NR1. Demolished *c.* 1895–1900.

Built by Henry Lock in 1829, Bracondale Mill was another nine-storey titan. Like all but one of Norwich's many mills, no trace remains as it disappeared from the city's skyline long ago. Henry Lock put Bracondale Mill up for auction in 1835 along with his mill at Poringland, which was constructed in 1825 but later demolished. Having built a number of mills (including Lakenham Peafield – see Chapter 1) in a short space of time, Mr Lock encountered financial problems. Between 1839 and 1865 the mill was owned by T. W. Read & Co. It was struck by lightning on 8 May 1872, but continued to operate after undergoing repairs. The mill was bought by Jeremiah James Colman and Frederick Edward Colman of Colman's Mustard fame in September 1877, before being dismantled at some point between about 1895 and 1900 by millwright Thomas Cook.

The tower was reported to measure 40 feet (12.19 metres) in diameter at the base – not far off the claims made for Bixley Mill and Southtown High Mill. The two early images reproduced here give a glimpse of a bygone age when the horse and cart was still the main form of transport.

Lakenham Bracondale Mill, *c.* 1855. (J. Neville)

Lakenham Bracondale Mill, *c.* 1855.
(J. Neville)

Watlington Tower Mill near King's Lynn

Location: Mill Road, Watlington, PE33 0HJ. Approx. 6 miles south of King's Lynn.
Demolished 1927/8.

Watlington Tower Mill stood eight storeys high and was one of west Norfolk's
tallest mills. It had an ogee (onion-shaped) cap complete with a gallery and topped
with a ball finial. A six-bladed fantail was fitted and the tower had a reefing stage at
fourth floor level. The four double-shuttered patent sails were each reported to be
9 feet (2.74 metres) wide with a large combined diameter of 91 feet (27.74 metres).
By comparison, the largest sail diameter of any surviving British Mill is 85 feet
(25.9 metres) (see Berney Arms Mill, Chapter 2).

Watlington Tower Mill *c.* 1912. (J. Neville)

Watlington Mill was built in 1847 and a steam engine was later installed to provide additional power. A workman was reported to have been killed by a falling chimney in June 1889 while attempting to install new ovens in the onsite bakery. The windmill was repaired following gale damage in February 1908 and finally stopped work in 1920. During much of its life it was owned by members of the Heading family and was worked by many different millers. Other owners included Edmund Crowe and Alfred Turner. The top of the mill was pulled off using ropes and chains before the brick tower was dismantled by hand during 1927/8.

Terrington St Clement – Balsam Fields Mill (aka Walker's Mill)

Location: Balsam Fields, Station Road, Terrington St Clement, PE34 4PL. Off A17, approx. 7 miles west of King's Lynn. Converted remains privately owned and not open to the public. Truncated 1908.

Standing on the edge of the Fens in West Norfolk, Balsam Fields Tower Mill in Terrington St Clement was one of the county's best known and tallest six-sailers. Formerly a seven-storey tower it sadly joined the ranks of Norfolk's many truncated windmills long ago. As with Bixley Mill the converted remains are on private property and are not accessible to the public. In its working days it had an ogee cap with gallery, petticoat and a ball finial, plus an eight-bladed fantail. The diameter of the sails was around 74 feet (22.5 metres). The tower had a reefing stage on the fourth floor and was 70 feet (21.3 metres) high to the top of the brickwork, making it taller than some eight or nine storey mills. A total height of well over 80 feet (24.38 metres) to the top of the finial is likely. The internal diameter at the base is approximately 25 feet (7.62 metres) and the thickness of the brickwork is 27 inches (0.68 metres).

Balsam Fields (aka Walker's) Mill, Terrington St. Clement, c. 1890. (J. Neville)

Alternatively known as Walker's Mill after Arthur Walker, who was the occupier in the 1880s, it was erected before 1841 by millwrights Eastwicks of Kings Lynn. It housed four pairs of millstones and was claimed to have been one of the most powerful windmills in Norfolk. In addition to all this wind power, a 10 hp engine was said to be driving two more pairs of stones by 1846. The mill's original owner was William Wright Snr whose son – also named William – was its first miller. John Newcome Wright Snr was the owner and miller between 1863 until his death in 1875. John Newcome Wright Jnr later became owner and he was also the miller from about 1896–1900. Several other millers came and went over the years, the last of the line being Robert Fordham (*c*. 1900–03). The windmill was damaged by a gale in March 1895 and finally stopped work in 1903. The top was blown off during another severe gale in February 1908, but by this time work was already in progress to partially dismantle it. After being stripped of its machinery the mill was reduced to three storeys and a height of about 30 feet (9.1 metres), before being topped with a conical corrugated iron roof. The present-day 'stump' that was once one of Norfolk's most impressive windmills still manages to stand a little taller than several other buildings on the site – most notably two former granaries – but gives little clue to its former existence as one of the Fen Country's most prominent local landmarks.

Another six-sailed mill existed at Orange Farm, Terrington St Clement, but was dismantled in 1920. Robert Fordham was previously at Orange Farm Mill before moving to Balsam Fields.

Diss – Victoria Road Tower Mill (aka Button's Mill)
Location: Victoria Road, Diss, IP22. On the south side of the A1066. Converted, remains privately owned and not open to the public.

This large mill is thought to have had eight sails when it was constructed around 1817 on what was then Diss Common. It was probably converted to a conventional four-sailer following severe gale damage in late 1836. It may have been at this time that it gained a huge dome-shaped cap 22 feet (6.7 metres) in diameter. Though not the tallest, Victoria Road Mill was also notable for its wide six-storey tower measuring 33 feet (10 metres) in diameter at the base, and a massive wooden brake wheel 11.5 feet (3.5 metres) in diameter. Four pairs of French burr stones were wind-driven and a further pair of stones was powered by a steam engine by 1880. This was later superseded by an oil engine. The mill's long association with the Button family began in 1880 when John Button became its new owner. It then stayed with the same family for the rest of its working life. After being reduced to one pair of sails in 1928 it carried on for about a year-or-so before work finally ceased. The tower was later reduced to four storeys and became part of a saw mill. Victoria Road Mill was in a derelict state when Mr Robert Manning purchased it in 1968 but he converted it to residential use and lived there for about thirty-five years. New windows were installed around the top of the truncated tower and the exterior was cement rendered. The new light colour scheme along with the other changes made it almost unrecognisable as the former big black mill with the impressive cap. Over the years the condition of the mill and the old mill house gradually deteriorated and they were both unoccupied when advertised for sale at the end of 2011.

Victoria Road (aka Button's) Mill, Diss, *c.* 1910. (M. Roots)

Bibliography

BOOKS

Apling, Harry, *Norfolk Corn and Other Industrial Windmills Volume 1* (Norwich: Norfolk Windmills Trust, 1984).

Ashley, Peter, *Up in the Wind* (Swindon: English Heritage, 2004).

Bonwick, Luke, *Norfolk's Windmills by River, Road and Rail* (Bonwick Publishing, 2008).

Brown, R.J., *Windmills of England* (Robert Hale & Co., 1976).

Flint, Brian, *Windmills of East Anglia* (Ipswich: F.W. Pawsey & Sons, 1971).

Hutchinson, Sheila, *Berney Arms Remembered* (Norwich: S. & P. Hutchinson, 2003).

Hutchinson, Sheila, *Reedham Memories* (Norwich: S. & P. Hutchinson, 2007).

Malster, Robert, *The Broads* (Chichester: Phillimore & Co. Ltd., 1993).

Page, Mike and Alison Yardy, *A–Z of Norfolk Windmills* (Wellington: Halsgrove, 2011)

Wailes, Rex, *Berney Arms Windmill* (HMSO, 1st edition 1957, 2nd edition 1982, 3rd edition English Heritage, 1990).

USEFUL WEBSITES

Norfolk Mills	www.norfolkmills.co.uk
The Mills Archive	www.millsarchivetrust.org
S.P.A.B. Mills Section	www.spab.org.uk/mills
The Norfolk Windmills Trust	www.norfolkwindmills.co.uk
Windmill World	www.windmillworld.com
UK Mills	www.ukmills.co.uk

Acknowledgements

All photographs taken by and copyright of the author except where stated.
Sutton Mill 1937 (two photographs) and Ingham Mill 1937: original unpublished photographs supplied by John Middleton from his family collection.
Polkey's Mill 1899 & Cadge's Mill 1930s: supplied by Peter Allard from his collection.
Garboldisham Mill March 2011: copyright of Adrian Colman and used with permission.
Upper Hellesdon Mill c. 1906: Previously published by John Nickalls Publications in a book of old picture postcards by Rhoda Bunn. Used with permission.
William & Susannah Hewitt, Berney Arms c. 1905.
From the author's own family collection.
Various additional photographs supplied by Jonathan Neville.

Thanks to the following people for their assistance:

John and Rosalind Middleton, Paul and Sheila Hutchinson, Jonathan Neville (Norfolk Mills website), Adrian Colman, Peter Allard, Debra Nicholson (curator of Wind Energy Museum), Mark Nickalls, Christine Bunn.